LIBRARY OF
NATURAL
DISASTERS

I0067425

HURRICANES,
TYPHOONS, & OTHER TROPICAL CYCLONES

WORLD
BOOK

www.worldbook.com

World Book, Inc.
180 N. LaSalle Street, Suite 900
Chicago, IL 60601
USA

For information about other World Book publications, visit our website at
http://www.worldbook.com or call **1-800-WORLDBK (967-5325)**.

For information about sales to schools and libraries, call
1-800-975-3250 (United States); 1-800-837-5365 (Canada).

3rd edition

© 2018 World Book, Inc. All rights reserved. This book may not be reproduced in whole or in part in any form without prior written permission from the publisher.

WORLD BOOK and the GLOBE DEVICE are registered trademarks or trademarks of World Book, Inc.

Library of Congress Cataloging-in-Publication Data

Title: Hurricanes, typhoons, & other tropical cyclones.
Description: 3rd edition. | Chicago, IL: World Book, Inc., [2018] | Series:
    World Book's library of natural disasters | Includes index.
Identifiers: LCCN 2017054261
    ISBN 9780716699354 (hc.) | ISBN 9780716694816 (pf.)
Subjects: LCSH: Hurricanes--Juvenile literature. | Typhoons--Juvenile
    literature. | Cyclones--Juvenile literature. | Natural disasters--Juvenile
    literature.
Classification: LCC QC944.2 .H87 2018 | DDC 551.55/2--dc23 LC record available
at https://lccn.loc.gov/2017054261

Set: ISBN: 978-0-7166-9928-6 (hc.)

Editor in Chief: Paul A. Kobasa

Print Content Development

   Director: Tom Evans
   Managing Editor: Jeff De La Rosa

Editors: William D. Adams,
   Nicholas Kilzer

Researcher: Jacqueline Jasek

Manager, Contracts & Compliance
(Rights & Permissions):
   Loranne K. Shields

Manager, Indexing Services:
   David Pofelski

Graphics and Design

   Coordinator, Design Development
   and Production: Brenda B. Tropinski
   Senior Visual Communications
   Designer: Melanie Bender
   Media Editor: Rosalia Bledsoe
   Senior Cartographer: John M. Rejba

Production

   Manufacturing Manager:
   Anne Fritzinger
   Proofreader: Nathalie Strassheim

Product development:
Arcturus Publishing Limited

   Writer: Neil Morris
   Editors: Nicola Barber, Alex Woolf
   Designer: Jane Hawkins
   Illustrator: Stefan Chabluk

Acknowledgments:

All maps and illustrations were prepared by the World Book staff unless otherwise noted.

AP Photo: 26 (Jose Luis Magana), 30, 36 (Xinhua/Liu Fei), 42 (PTI), 43 (Sipa Press), 45, 48.

The Art Archive: 6 (Laurie Platt Winfrey), 7 (Society of Apothecaries/ Eileen Tweedy).

Corbis: 10 (Cuartoscuro/ R.V. Altierra/ Corbis Sygma), 14 (Imelda Medina/ epa), 17 (NOAA), 19 (Jim Reed),
    20 (Reuters), 21 (Corbis), 23 (David J. Phillip/ epa), 25 (Michael Ainsworth/ Dallas Morning News), 34, 35, 47
    (Bettmann), 37 (Jiang Kehong/ Xinhua Press), 39 (Les Stone/ Sygma), 44 (Christian Simonpietri/ Sygma).

Federal Emergency Management Agency: 29 (Jocelyn Augustino).

Getty Images: 24 (Mark Wilson), 29 (Lionel Chamoiseau).

NASA: cover (Image Science and Analysis Laboratory), 4 (Jesse Allen/ Earth Observatory, using data provided
    courtesy of the MODIS Rapid Response), 32.

National Oceanic and Atmospheric Administration: 15, 27, 41 (Department of Commerce).

Science Photo Library: 8, 50, 51 (Jim Reed), 33 (Jim Edds).

Sipa Press: 22 (Thomas Haley).

U.S. Air National Guard: 28 (Staff Sgt. Daniel J. Martinez).

TABLE OF CONTENTS

Glossary There is a glossary of terms on pages 53-54. Terms defined in the glossary are in type **that looks like this** on their first appearance on any spread (two facing pages).

Additional resources Books for further reading and recommended websites are listed on page 55. Because of the nature of the Internet, some website addresses may have changed since publication. The publisher has no responsibility for any such changes or for the content of cited sources.

WIND AND CYCLONES

Wind is air that moves across the surface of Earth. Sometimes the wind is a gentle breeze that barely moves the leaves on the trees. At other times, the wind blows hard enough to cause great damage to trees, buildings, and other structures. Sometimes winds bring together masses of air that are at different temperatures and that contain different amounts of **humidity.** When these different air masses meet, they can form a **weather system** called a **cyclone** *(SY klohn).*

A satellite image shows Hurricane Ileana in August 2006. The violent storm took several days to move northward along the Pacific coast of Mexico.

Weather system

The word *cyclone* is often used to mean a violent, swirling windstorm. Another meaning for the term refers to the weather system in which these windstorms form. In such systems, a warm air mass rises and is replaced by cooler air, creating an area of low **atmospheric** *(AT muh SFEHR ihk)* **pressure,** or air pressure, near Earth's surface. A cyclone usually produces clouds and rain, as well as strong winds that spin around the area of low pressure. These weather systems occur north of the **tropics** in the **Northern Hemisphere** and south of the tropics in the **Southern Hemisphere.** They can affect the weather over a very large area of land and sea.

In the tropics

A more violent type of cyclone develops in the warm regions known as the tropics. **Tropical cyclones** form in humid air over warm ocean waters. The powerful winds of a tropical cyclone spin around a calm center, called the **eye.** Tropical cyclones are often extremely violent storms.

Around the world

Tropical cyclones are known by various terms in different parts of the world. They are called **hurricanes** when they occur over the North Atlantic Ocean, including the Caribbean Sea and the Gulf of Mexico, or over the Northeast Pacific Ocean. They are called **typhoons** *(ty FOONz)* in the Northwest Pacific Ocean. In the Indian Ocean and the South Pacific Ocean near Australia, they are normally referred to as tropical cyclones (or sometimes simply cyclones). In this book, the word *hurricane* is used as the general term for a tropical cyclone.

NAMING TROPICAL CYCLONES

Hurricanes, typhoons, and other tropical cyclones are given individual names to identify them. This system helps to avoid confusion when there is more than one severe storm occurring at the same time or when discussing multiple storms that happened in the same year. The World Meteorological Organization, based in Switzerland, helps regional committees around the world choose the names. Each year, Atlantic hurricanes start with a name that begins with the letter A and continue through the alphabet, alternating between female and male names. In 2017, the names chosen for the first five tropical storms over the Atlantic were Arlene, Bret, Cindy, Don, and Emily.

If a hurricane is especially devastating, that hurricane's name will usually not be used again. This prevents confusion over which Hurricane Rita, for example, is being discussed. In 2005, a number of deadly hurricanes occurred, so the names Dennis, Katrina, Rita, Stan, and Wilma were all *retired* (taken out of use).

Where hurricanes, typhoons, and tropical cyclones occur

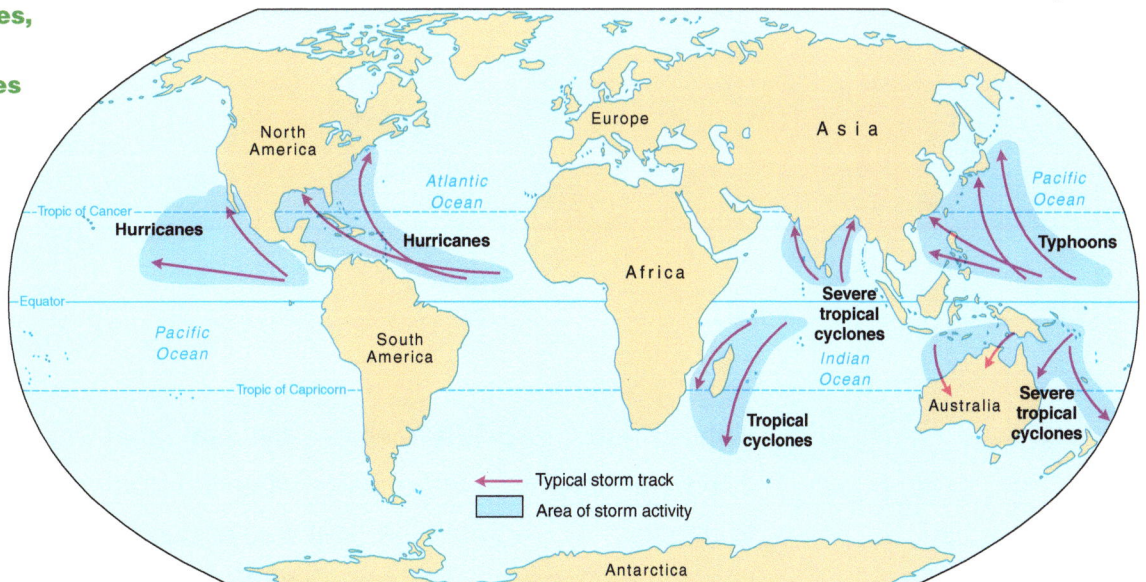

CHANGING HISTORY

Damaging storms have caused devastation around the world since prehistoric times. **Hurricanes** were certainly a threat to people in ancient times. In fact, scientists studying the American Gulf Coast from Florida to Texas have found evidence that catastrophic hurricanes struck the region more frequently 3,000 years ago than they do today.

Storm gods

Some ancient civilizations had storm gods. The Taíno of the Caribbean and the Maya of Mexico and Central America believed that the god Hurakán caused winds and storms. According to their legends, it was Hurakán who showed the gods' displeasure with humans by sending a great flood. The name Hurakán may be the origin of the English word *hurricane.* The word came into usage in English after Spanish explorers in the Americas in the 1500's adopted the word *huracán* to describe a severe storm.

Divine winds

Sometimes, a tropical cyclone has actually been of help to some people. In the late 1200's, **typhoons** prevented Japan from being invaded. The Mongol leader Kublai Khan (1215-1294), whose dynasty ruled China from 1279 to 1368, put together a large fleet that sailed from the Korean peninsula to Japan in 1274. While the fleet was anchored off the Japanese coast, a typhoon wrecked one-third of the fleet's ships and left about

Kublai Khan's fleet, attempting to invade Japan in 1274, struggles to stay afloat during a typhoon, from a Japanese scroll of the time. The storm sank one-third of the fleet, saving Japan from being invaded.

13,000 Mongol sailors dead, saving Japan from invasion.

In 1281, the Mongols tried again. But after weeks of battle off the Japanese coast, another typhoon drove the ships ashore, killing most of the 140,000 invaders. The Japanese honored their good fortune by naming the typhoon *kamikaze (KAH mee KAH zee),* which means "divine wind."

During World War II (1939-1945), the term kamikaze was used to describe Japanese suicide pilots. These pilots were thought by the Japanese to be like a typhoon—raining down destruction. Kamikaze pilots deliberately crashed their planes into U.S. warships, often causing major damage and loss of life.

SPANISH ARMADA

In 1588, King Philip II (1527-1598) of Spain sent a fleet known as the Spanish Armada *(ahr MAH duh)* to invade England. In the English Channel, the Armada was attacked by the English fleet and forced to flee northward around Scotland. The Armada then turned south and sailed along the Irish coast, where hurricane-force winds sank some ships and drove others onto rocks. Only two-thirds of the original Armada eventually returned to Spain.

After being outmaneuvered by the English fleet in 1588, the Spanish Armada was further devastated by hurricane-force winds off the coast of Ireland.

BUILDING UP TO A HURRICANE

Hurricanes form over seas and oceans with a surface temperature greater than 80 °F (27 °C). In such waters, the seawater **evaporates** into the air above it, forming a rising column of warm, moist air. This creates an area of low **atmospheric pressure** just above the water's surface, and because air tends to move from areas of high pressure to low pressure, it also creates wind.

Tropical disturbance

Meteorologists (*MEE tee uh ROL uh jihstz*) call the first stage in the life of a hurricane a *tropical disturbance.* As the warm, moist air rises off the sea, it starts to cool down. Because cool air cannot hold as much **water vapor** as warm air, the vapor changes back

A fisherman struggles to secure his boat to a dock in Gulfport, Mississippi, in the southern United States, as a tropical storm approaches.

into droplets of water and forms clouds. These become huge, towering thunderclouds, which meteorologists call **cumulonimbus** (KYOO *myuh loh NIHM buhs) clouds.* As long as there is enough warmth and moisture in the atmosphere, the cumulonimbus clouds continue to form.

Tropical depression

The second stage in the formation of a hurricane is called a tropical **depression.** This is an area of low atmospheric pressure that is large enough to be plotted on a weather map. The region of low pressure near the warm ocean continues to draw in warm, moist air, creating winds that blow in a circular pattern. As the pressure drops even lower, more and more warm air flows in, and the winds move faster. The storm is now well on the way to becoming a hurricane.

In the first barometers, a glass tube in a basin of mercury stood upright against a scale.

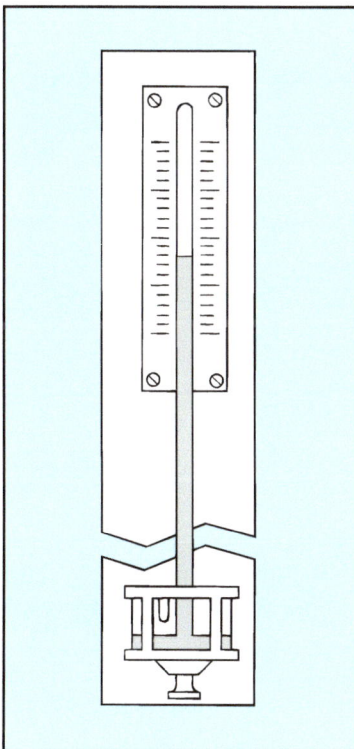

MEASURING ATMOSPHERIC PRESSURE

Atmospheric pressure is measured using an instrument called a **barometer.** (See page 52 for instructions on making your own barometer.) The Italian scientist Evangelista Torricelli (*eh* VAHN *jeh LEE stah* TAWR *ih CHEHL ee;* 1608-1647) introduced the first barometer in 1644. He was trying to solve a practical problem—the inability of vacuum pumps to raise water higher than about 34 feet (10 meters). Torricelli suspected the reason might be atmospheric pressure, and he began experimenting. He filled a glass tube with **mercury,** put the sealed end of the tube facing up, and then he stood the tube with the open end down in a basin of mercury. The column of mercury in the tube fell until its top was about 30 inches (76 centimeters) above the surface of the mercury in the basin. The pressure of the air on the surface of the liquid in the basin was holding the remaining mercury in the tube. As the weight of the air, or the atmospheric pressure, decreased or increased, the mercury level in the tube rose and fell.

Atmospheric pressure may be measured in several ways. One way is in units called **millibars (mb).** The average pressure at sea level is 1,013 mb. In 1979, **Super Typhoon** Tip set a record low for atmospheric pressure at 870 mb.

Violent, whirling winds

When the winds in a tropical **depression** blow at more than 38 miles (61 kilometers) per hour, the third stage of hurricane formation is reached. It is then officially a tropical storm. A tropical storm's clouds have a well-defined circular shape. Its strong winds near the ocean surface draw in increasing amounts of heat and **water vapor.** As the column of air near the center of the tropical storm gets warmer, pressure continues to fall, and the storm gets stronger. Each tropical storm is given a name, such as Tropical Storm Andrea. Should a tropical storm become a **hurricane,** it keeps its name but changes its designation—that is, Tropical Storm Andrea becomes Hurricane Andrea.

Final stage

A tropical depression becomes a hurricane when its winds accelerate to 74 miles (119 kilometers) per hour or more. By this time, the storm has a well-developed **eye** at its center (see page 16). The smaller

Vehicles lie half buried in the sand on a beach near Acapulco, Mexico, in the aftermath of Hurricane Pauline. The 1997 storm caused severe flooding in Acapulco, killing some 200 people.

the eye, the faster the winds blow around it.

As a hurricane continues to draw heat and moisture from the sea, it becomes ever more violent. In 1972, U.S. civil engineer Herbert S. Saffir and **meteorologist** Robert H. Simpson devised a scale, called the Saffir-Simpson hurricane scale, to indicate the intensity of a hurricane.

The Saffir-Simpson hurricane scale

The Saffir-Simpson scale is based on wind speed in miles and kilometers per hour and the height of the resulting **storm surge**— that is, how much the sea level rises above normal high **tide.**

COLUMBUS'S HURRICANE

In 1502, Christopher Columbus met with hurricane winds on the coast of Panama during his fourth voyage to the Americas. He survived to describe the storm: "The tempest arose and wearied me so that I knew not where to turn. ... Eyes never beheld seas so high, angry, and covered by foam. The wind not only prevented our progress, but offered no opportunity to run behind any headland for shelter. ... Never did the sky look more terrible; for one whole day and night it blazed like a furnace, and the lightning broke forth with such violence that each time I wondered if it had carried off my spars and sails. ... All this time the water never ceased to fall from the sky; I don't say it rained, because it was like another deluge."

Hurricane category	Wind speed		Effects
	mph	kph	
1 (weak)	74-95	119-153	Minimal damage to trees, shrubbery, and mobile homes. Storm surge could be 4-5 feet (1.2-1.5 meters) above normal water level.
2 (moderate)	96-110	154-177	Considerable damage to trees, mobile homes, and piers; some damage to roofs. Storm surge could be 6-8 feet (1.8-2.4 meters) above normal water level.
3 (strong)	111-130	178-209	Trees blown down or stripped of leaves; mobile homes destroyed; some damage to other buildings. Storm surge could be 9-12 feet (2.7-3.7 meters) above normal water level.
4 (very strong)	131-155	210-249	Extensive damage to windows, doors, and roofs, especially near shore; possible flooding. Storm surge could be 13-18 feet (4.0-5.5 meters) above normal water level.
5 (devastating)	156 and up	250 and up	Small buildings overturned or blown away; severe structural damage to other buildings. Storm surge could be 19 feet (5.8 meters) above normal water level, or higher.

HURRICANE MITCH

Atlantic hurricanes

Scientists have discovered that many Atlantic **hurricanes** develop from areas of low **atmospheric pressure** over tropical West Africa. These storms develop as they move west off the continent into the Atlantic Ocean. When this happened in October 1998, the result was Hurricane Mitch—the deadliest hurricane to strike the **Western Hemisphere** since the Great Hurricane of 1780 (see pages 14-15).

From depression to storm to hurricane

An area of low pressure moved off the African coast on Oct. 10, 1998. It took eight days to move across the Atlantic Ocean. By October 22, the region of low pressure had become a tropical depression—the 13th tropical depression of the Atlantic hurricane season, which runs from June to November. The depression had wind speeds of 35 miles (56 kilometers) per hour. Drifting westward, it grew into a tropical storm later that same day and was named Mitch. The storm became a hurricane on October 24, when it was 293 miles (472 kilometers) south of Jamaica. By October 26, the hurricane's atmospheric pressure

The path taken in 1998 by Hurricane Mitch across Central America

Dates show positions of Hurricane "Mitch" in October - November 1998

★ National capital

0   200 Miles

0   200 Kilometers

had dropped to 905 **mb,** and the wind speed had grown to 180 miles (290 kilometers) per hour. The hurricane, now at a devastating Category 5, headed west toward Honduras, in Central America.

Devastation

By the time Hurricane Mitch made **landfall** on October 29, its wind speeds had weakened somewhat—but it was still awesome in its strength. On its way to Honduras, Mitch drove waves to heights of up to 44 feet (13 meters), flooding the Honduran coast. As the storm moved across the mountains of Honduras and headed toward Guatemala, it released enormous amounts of rainfall. The subsequent floods and **mudslides** devastated Honduras and parts of Nicaragua, Guatemala, Belize, and El Salvador.

Hurricane Mitch killed at least 11,000 people and left about 1.5 million people homeless. People were still clinging to rooftops a week after the storm, waiting to be rescued. By then, the weakened hurricane was making landfall near Naples, Florida, still with wind gusts of 80 miles (129 kilometers) per hour. The name Mitch was retired after this hurricane.

A LOST SHIP

Fantome is French for *ghost*. The yacht renamed *Fantome* in 1937 was built for a member of the British aristocracy in 1927, as the *Flying Cloud*. Later, the wealthy Greek businessman Aristotle Onassis bought this beautifully appointed yacht as a wedding present for Prince Rainier and Princess Grace of Monaco. Onassis never gave it, however, because he was not invited to the wedding. In 1969, the *Fantome* was purchased again, given four masts and a steel hull, and became a cruise ship in the Caribbean. On Oct. 24, 1998, the ship loaded with passengers and left from Honduras on a cruise to Belize.

Weather forecasters began predicting that Hurricane Mitch would be moving toward Belize, so the captain of the *Fantome* decided to cruise around the area of Honduras. The forecasting models were wrong this time. As Mitch became more violent, the ship's captain decided to drop all of the passengers and a few crew members off at Belize City, using launches to get the people from the ship to the shore. Then, he tried to get the *Fantome* away from the dangerous shorelines, hoping to ride out the storm in the Gulf of Mexico.

Because Mitch did not follow the path that forecasters had predicted, the *Fantome* was actually sailing for the center of the storm. During the ship's last transmission on October 27, the crew reported winds at speeds of 100 miles (160 kilometers) per hour and 40-foot (12-meter) waves. No one knows what happened next, but the 282-foot (86-meter) ship was lost with all remaining 31 members of its crew.

THE GREAT HURRICANE OF 1780

In October 1780, an Atlantic **hurricane** struck various islands of the Caribbean. Historians believe that the Great Hurricane of 1780 left about 22,000 people dead—the highest death toll for any Atlantic hurricane on record. The dead included about 9,000 people on the French island of Martinique, more than 9,300 people on British Barbados, at least 9,000 on the Dutch island of Saint Eustatius, and about 6,000 people on the British island of Saint Lucia. The storm also killed thousands of French and British sailors who were engaged in a battle for control of the Caribbean.

In fact, the entire hurricane season of 1780 was incredibly devastating. Over the course of the season, eight separate storms hit the coasts of the West Indies and Americas. The number of people killed by these storms is thought to be around 25,000.

Out of the blue

When Hurricane Mitch struck 218 years later, weather forecasters could at least predict its likely impact, if not its course. But in 1780, the science of **meteorology** (MEE tee uh ROL uh jee) was still in its infancy. Although sailors were on the lookout for

The Caribbean experiences frequent hurricanes.

Not even the most intense Caribbean storms over the last century have been as devastating as the Great Hurricane of 1780.

signs of bad weather, the Great Hurricane probably struck with little warning. There was no chance to prepare for it, and no way of sending advance warning to the people on other islands in the hurricane's path.

The first island to be hit was Barbados, on October 10. The hurricane damaged and scattered the British fleet off the island of Saint Lucia before heading on to Martinique, where it sank more than 40 ships, carrying 4,000 soldiers. The hurricane destroyed crops throughout the region, and thousands of people later died from famine. Food supplies could not be imported to replace the lost crops because trade was disrupted by the Revolutionary War in America, which was being fought at the time.

CIRCULAR STORMS

Captain Henry Piddington (1797-1858), a British sailor stationed in India and an amateur meteorologist, published *Sailor's Horn-book for the Law of Storms* in 1848. In it he labeled violent, whirling storms as *cyclones.* He took the word from the Greek *kuklos,* meaning "coil of a snake." Piddington was the first person to describe tropical storms scientifically.

Eyewitness account

Admiral Lord George Rodney (1718-1792) was in the Caribbean in 1780, commanding part of the British fleet. On December 10, he wrote a letter to his wife describing the devastation on Barbados: "The strongest buildings and the whole of the houses, most of which were stone and remarkable for their solidity, gave way to the fury of the wind, and were torn up to their foundations; all the forts [were] destroyed, and many of the heavy cannon carried upwards of a hundred feet from the forts. Had I not been an eyewitness, nothing could have induced me to have believed it."

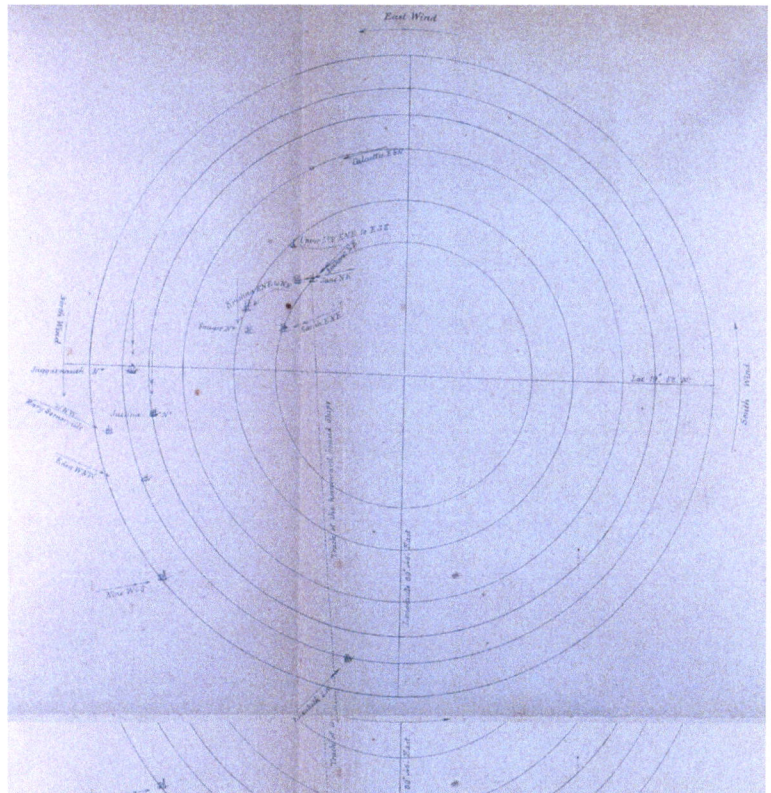

An 1839 diagram by Henry Piddington shows the circular nature of tropical cyclones.

THE EYE OF THE STORM

Henry Piddington (see page 15) described how the winds of a **tropical cyclone** spin around a "fatal centre." Today, this is called the **eye** of the storm. The eye is a relatively calm, funnel-shaped area of low **atmospheric pressure** that acts like the axle of a wheel. The rest of the **hurricane** spins around it. The eye of a hurricane is usually 10 to 40 miles (16 to 64 kilometers) in diameter. The winds created by the **eyewall** can reach nearly 200 miles (320 kilometers) per hour. Damaging winds may extend 250 miles (400 kilometers) from the eye.

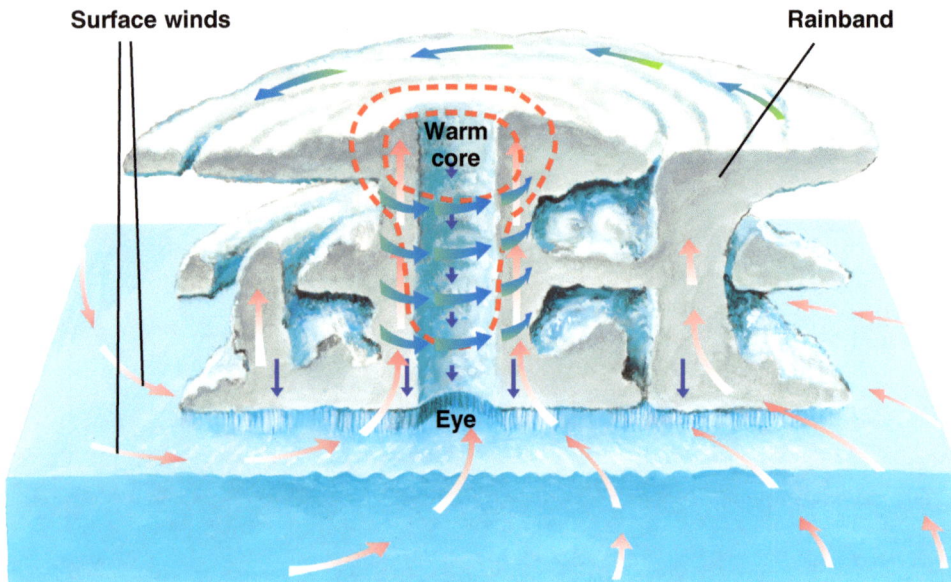

Surface winds

Rainband

Warm core

Eye

Surface winds feed heat energy and water vapor into a hurricane. A core of warm air maintains a zone of low pressure, drawing more air into the eye of the hurricane. Rain falls from clouds called the eyewall and rainbands. (Diagram is not to scale.)

Eyewall and rainbands

A wall of turbulent clouds, called the eyewall, surrounds the eye. In the eyewall, warm air spirals upward, creating the storm's strongest winds. The speed of the winds in the eyewall is related to the diameter of the eye. Just as ice skaters spin faster when they pull their arms in, a hurricane's winds blow faster if its eye is small. If the eye widens, the winds decrease.

Dense clouds swirl around the eyewall, and these spiral **rainbands** produce huge amounts of rain—sometimes more than 2 inches (5 centimeters) in an hour. The rainbands may stretch for hundreds of miles from the eyewall.

Clockwise or counterclockwise?

Hurricanes usually travel westward, pushed along by the tropical **trade winds**—steady winds that blow toward the **equator** from the northeast or southeast. Earth's rotation also affects the movement of hurricanes. In the **Northern Hemisphere,** Earth's rotation causes hurricanes to veer to the right and spin in a counterclockwise direction. In the **Southern Hemisphere,** they veer left and spin clockwise. The strength of the wind in a hurricane varies around the eye. An Atlantic hurricane (in the Northern Hemisphere) has stronger winds on its right side because these winds are moving in the same direction as the steering wind. The left-hand winds are blowing against the wind that is moving the storm along, which weakens the winds on that side.

HURRICANE HUNTERS

To predict the course and intensity of a hurricane, **meteorologists** need to know the exact location of the pressure center of the storm's eye and wind speeds in the eyewall. In the United States, the specially trained crews of the 53rd Weather Reconnaissance Squadron, part of the U.S. Air Force Reserve, fly weather-related missions during the hurricane season. This squadron is commonly known as the Hurricane Hunters. They fly into the eye of a storm to get the information meteorologists need for prediction purposes.

You might think that planes would break apart flying into a hurricane. But, planes are designed to fly through wind. Under normal conditions at higher altitudes, wind speeds can reach more than 200 miles (320 kilometers) per hour, which is higher than the winds of all but the worst of Category 5 hurricanes. Hurricanes are actually much harder on objects that cannot move, such as houses or trees. When a plane is hit by a hard gust of wind, it may jolt up or down, and the ride may be bumpy, but the plane can move with the wind. Stationary objects cannot, and so they are often damaged by intense winds.

The eyewall of Hurricane Katrina, is shown in an image taken from a Hurricane Hunter aircraft. The plane flew right through the storm.

STORM SURGES

The most dangerous effect of a **hurricane** is a rapid rise in sea level. This is called a **storm surge.** According to the U.S. National Oceanic and Atmospheric Administration (NOAA), storm surges have been responsible for 9 out of every 10 deaths caused by hurricanes since records have been kept. The Saffir-Simpson hurricane scale relates the height of a storm surge to wind speed (see page 11).

Whipping up the waves

Violent hurricane winds whip up the sea beneath them and create huge waves. At the center of the storm, the low **atmospheric pressure** in the **eye** pulls up a bulge of water. At the same time, the entire circular storm is moving, usually at a speed of 10 to 20 miles (16 to 32 kilometers) per hour. As the hurricane nears a coastline, its winds drive the ocean waters ashore. Often huge waves begin to crash on the shoreline well before the most violent winds arrive. If the eye of the hurricane moves over the coast, the bulge of water beneath it may also contribute to the flooding.

A storm surge adds to the height of normal tides, creating a storm tide. As the hurricane winds increase, huge waves are formed on top of the storm tide.

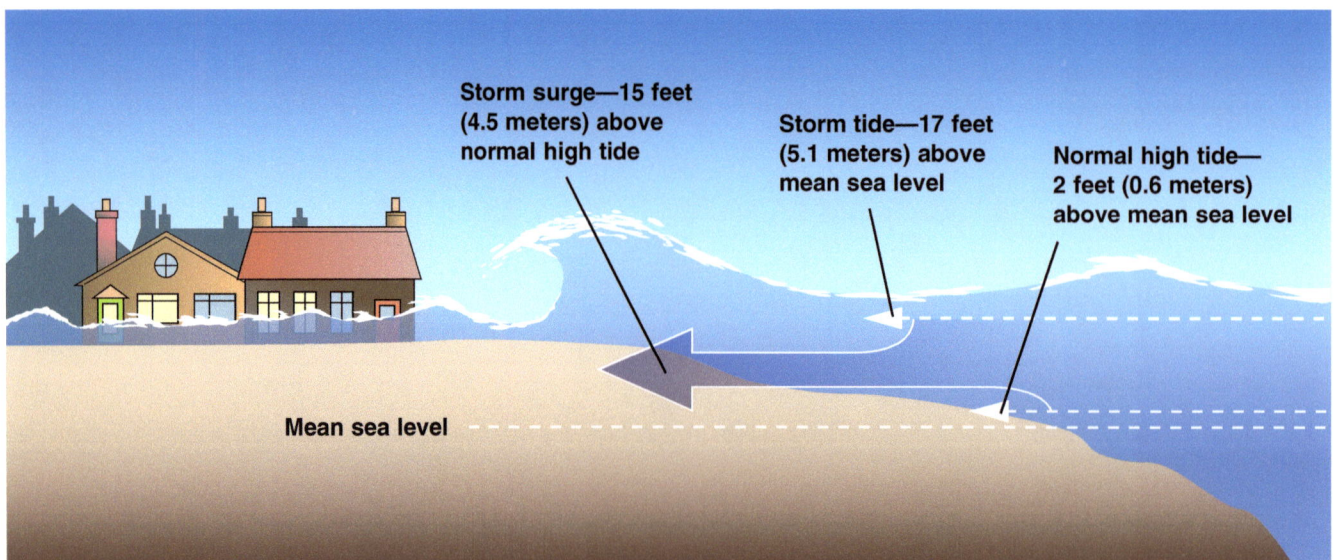

Storm surge—15 feet (4.5 meters) above normal high tide

Storm tide—17 feet (5.1 meters) above mean sea level

Normal high tide— 2 feet (0.6 meters) above mean sea level

Mean sea level

Flood risk

Storm surges can cause severe flooding. If the surge arrives at the coast at the same time as a normal high **tide,** it can produce a **storm tide** resulting in devastating floods. This is especially true if the coast has a shallow slope, rather than a steep incline or cliff face, because the force of the surge can drive the water far inland. Many coastal cities in hurricane zones around the world are only a few feet above sea level—some, such as New Orleans, are actually below sea level. These cities are very vulnerable to storm surges.

A truck driving on the oceanfront highway on Hatteras Island off North Carolina in the southern United States is caught in a hurricane storm surge. The storm surge delivered by Hurricane Isabel in 2003 destroyed whole sections of the highway.

DEADLY STORM SURGES

The most deadly natural disaster in U.S. history was caused by a storm surge. Galveston, Texas, (see pages 20-21) is built on a barrier island in the Gulf of Mexico. In 1900, the island was about 9 feet (3 meters) above sea level at its highest point. A storm surge caused by a hurricane sent waves as high as 15 feet (4.5 meters) sweeping over the entire island. There were so may dead on the island after the storm—either drowned or crushed by the waves and debris—that burial was impossible. Survivors tried to weight the bodies and bury them at sea, but they were washed back to shore. Eventually, piles of wood were set on fire and the bodies were cremated.

GALVESTON'S GREAT STORM

Debris from flattened buildings filled the street at Galveston, Texas, after the hurricane of 1900.

The route taken by the hurricane as it tracked across the Gulf of Mexico before hitting the city of Galveston in 1900

In the late 1800's, Galveston, Texas, was a major seaport on the Gulf of Mexico. The city lies on Galveston Island, about 2 miles (3 kilometers) off the Texas mainland. Founded in 1836, it was, by 1880, the largest city in Texas. In 1900, the city was virtually destroyed by the **storm surge** from a Category 4 **hurricane.** About 3,600 buildings were destroyed, and at least 6,000 of the city's 37,800 inhabitants died in the storm.

Sept. 8, 1900

The hurricane first hit Hispaniola—the island occupied by the nations of Haiti and the Dominican Republic—as a tropical storm on Sept. 1, 1900, before moving to Cuba and then into the Gulf of Mexico. On September 7, the chief of Galveston's weather station, Isaac Cline, hoisted a storm-warning flag. The following morning, Cline rode in a carriage along the beach and warned people to evacuate or seek shelter on higher ground, but most took little notice. And, to be fair, Cline had previously been quoted in the local newspaper saying that there was no possibility Galveston could be hit by a major hurricane. (Cline would survive the storm, along with some of his children, by floating for hours on **debris** and wreckage until the waters subsided.)

By early afternoon, the city was flooding and a huge wall of debris, from hundreds of buildings destroyed by the storm, was being pushed in by the rising **tide.** The **eye** of the hurricane passed just west of Galveston that evening, and by that time, the whole island was underwater.

The devastation after the storm was shocking. Bodies of the dead hung from trees and telegraph poles. Children from the local orphanage had been tied with clothesline to the nuns who ran the institution, in the hope of keeping the children close to an adult. They were all found dead on the beach, still bound together. Thousands of bodies floated near the shore.

The seawall in Galveston, pictured in 1909. The side of the wall facing the sea was sloped to lessen the impact of waves.

THE SEAWALL

The devastation of 1900 led the people of Galveston to build a protective seawall. Construction began two years later on a concrete wall 17 feet (5 meters) high. Once 3 miles (5 kilometers) long, it now stretches 10.4 miles (16.7 kilometers). Workers also pumped sand beneath houses before they were rebuilt, raising the city to a new height of 25.7 feet (7.8 meters) above sea level. Despite these efforts, Galveston never regained its importance as a hub of oil money and economic influence and was surpassed by Houston. The seawall proved only somewhat effective in preventing flooding. In 2008, another severe hurricane, Ike, hit Galveston. Water simply swept around the wall and flooded the city from the bay side.

HURRICANE KATRINA

Hurricane Katrina was one of the most destructive **hurricanes** ever to strike the United States. In late August 2005, Katrina brought high winds, huge waves, and flooding that caused damage in Florida and widespread destruction in Louisiana, Mississippi, and Alabama. The storm resulted in the deaths of more than 1,800 people and caused tens of billions of dollars in damage.

Areas of low pressure

Meteorologists note that Hurricane Katrina developed from a tropical **depression** that began on Aug. 23, 2005. The depression increased in strength as it moved northwest along the Bahamas. It became a Category 1 hurricane on August 25, just before striking the tip of Florida. The hurricane gained strength as it passed into the Gulf of Mexico. It made **landfall** near the Louisiana-Mississippi border on August 29 as a Category 3 hurricane. Wind gusts of

Flooding prevents firefighters from reaching a house fire in the French Quarter of New Orleans after Hurricane Katrina devastated the city. Fire was a major problem in the aftermath of the hurricane.

more than 125 miles (200 kilometers) per hour, together with **storm surges** reaching up to 29 feet (8.8 meters), devastated several coastal communities in Mississippi and Alabama.

Flooding of New Orleans

The **eye** of the storm made landfall on August 29 at Buras, Louisiana, about 50 miles (80 kilometers) east of the city of New Orleans. In the city, winds of up to 100 miles (160 kilometers) per hour damaged many buildings. But the worst destruction was caused by rapidly rising water levels. A storm surge pushed water over and broke through several parts of the New Orleans system of **levees** (flood barriers). The city's low-lying neighborhoods and eastern suburbs quickly flooded. Within 24 hours of the hurricane's striking the city, more than 80 percent of New Orleans was under as much as 20 feet (6 meters) of filthy, polluted water.

LEVEES

There have been levees in New Orleans since soon after the city was founded in 1718, and engineers have periodically added to the levees' height and length. They were originally built to keep the Mississippi River from flooding its banks. After floods caused by Hurricane Betsy in 1965, the city's system of levees was expanded and reinforced. But the walls were still not high or strong enough to withstand the storm surge from Hurricane Katrina. In some places, the water destroyed the levees; in others, water simply went over the top of the levee. The 2005 disaster led to a series of investigations aimed at improving flood defenses for New Orleans and other threatened cities.

A helicopter lowers sandbags into a breach in a New Orleans levee nine days after it failed in the wake of the Hurricane Katrina storm surge.

Hurricane Katrina destroyed a huge number of homes and businesses. Hundreds of thousands of people were suddenly left homeless. According to figures from the National Hurricane Center, 1,090 people were killed in Louisiana, 238 in Mississippi, 14 in Florida, 2 in Georgia, and 2 in Alabama. Most of the Louisianans who died were from New Orleans. The total cost of the damage caused by the storm was estimated at $125 billion.

Surviving the storm

The mayor of New Orleans ordered the evacuation of the city the day before the **hurricane** hit, but many people could not or did not want to leave their homes. Some people thought that the water level would soon go down again, and they were worried about other people breaking in and stealing. Others could not leave because they were poor and had no means of transportation or because they were too old or ill to travel. Helicopters and boats were still picking up survivors stranded on rooftops days after the

A military truck drives down a flooded street in New Orleans after Hurricane Katrina struck the city.

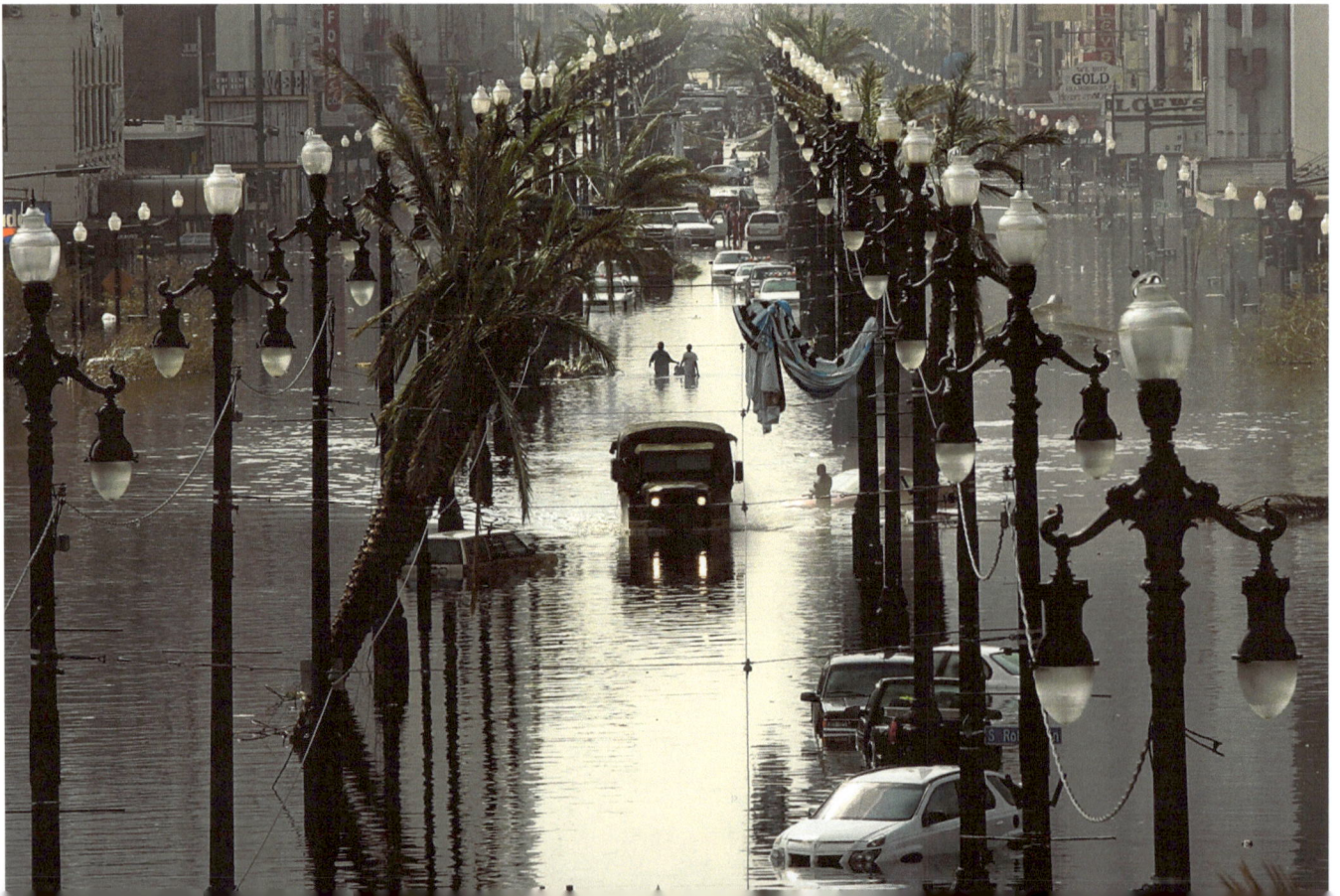

city flooded. On August 31—as the weakened storm was dying out near the Great Lakes region—up to 100,000 people remained in New Orleans. Thousands took shelter in the city's convention center and Superdome sports stadium, where conditions soon deteriorated. Some people who were sick or elderly were left in hospitals and nursing homes without drinking water or electricity, and many died. People waited for days for buses to take them to shelters in such cities as Baton Rouge, Louisiana, and Houston, Texas.

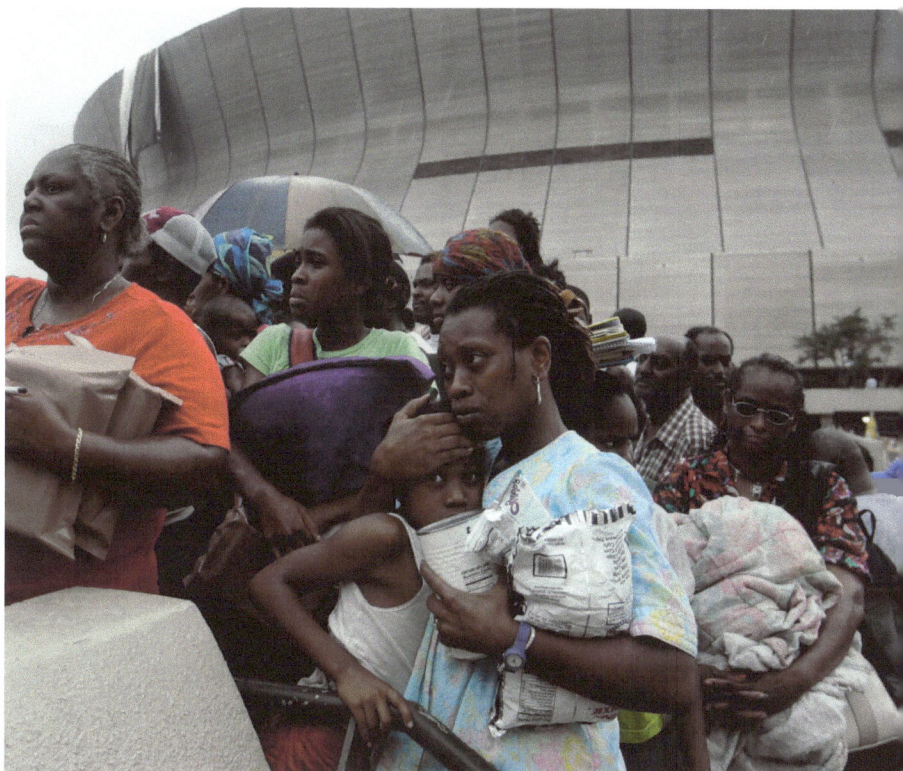

In the aftermath of Hurricane Katrina, in New Orleans, people wait outside the Superdome sports stadium for buses to take them out of the flooded city to emergency shelter in Houston.

Dealing with the emergency

On August 31, U.S. government officials declared a public health emergency along the whole Gulf Coast. People from around the world gave billions of dollars to charities to help the recovery operation. A week later, engineers started to pump water from New Orleans after repairing the main breaches in the levees. Later, there was criticism of the way the disaster was handled and an official investigation was begun. A year after Katrina, the levees had been repaired and rebuilt and there were plans to strengthen the defenses. These included the use of powerful pumps to help against future **storm surges** by moving water out quickly through the city's canals.

ONE YEAR LATER ...

New Orleans struggled to recover from the damage wrought by Katrina. The flooding destroyed as many as 275,000 houses. Twelve months later, only about half of the city's 485,000 residents had returned.

Areas that had been built on higher ground, such as the city's historic French Quarter, sustained less damage, but lower areas of the city, such as the Lower Ninth Ward, were devastated. Almost nothing had been rebuilt in the Ninth Ward one whole year after Katrina struck. The pollution in this area was so extreme that the order to boil all drinking water was not lifted until the spring of 2006.

HURRICANE WILMA

The most intense storm ever recorded in the Atlantic Basin occurred in 2005. **Hurricane** Wilma was a Category 5 storm that wreaked havoc on the Yucatán Peninsula, in Mexico, and south Florida, in the U.S., in October. The ferocious storm blew in just weeks after two other Category 5 storms—hurricanes Katrina and Rita—pounded New Orleans and other parts of the Gulf Coast region. Wilma, packing winds of 185 miles (300 kilometers) per hour, dumped more than 60 inches (150 centimeters) of rain on parts of the Yucatán. The storm killed nearly two dozen people in Haiti, Jamaica, Mexico, and Florida. It inflicted more than $29 billion in damage. Hurricane Wilma became the fourth costliest hurricane to hit the United States, after hurricanes Katrina, Sandy, and Ike.

Tourists caught by Hurricane Wilma sleep in an emergency shelter as the storm bears down on Mexico.

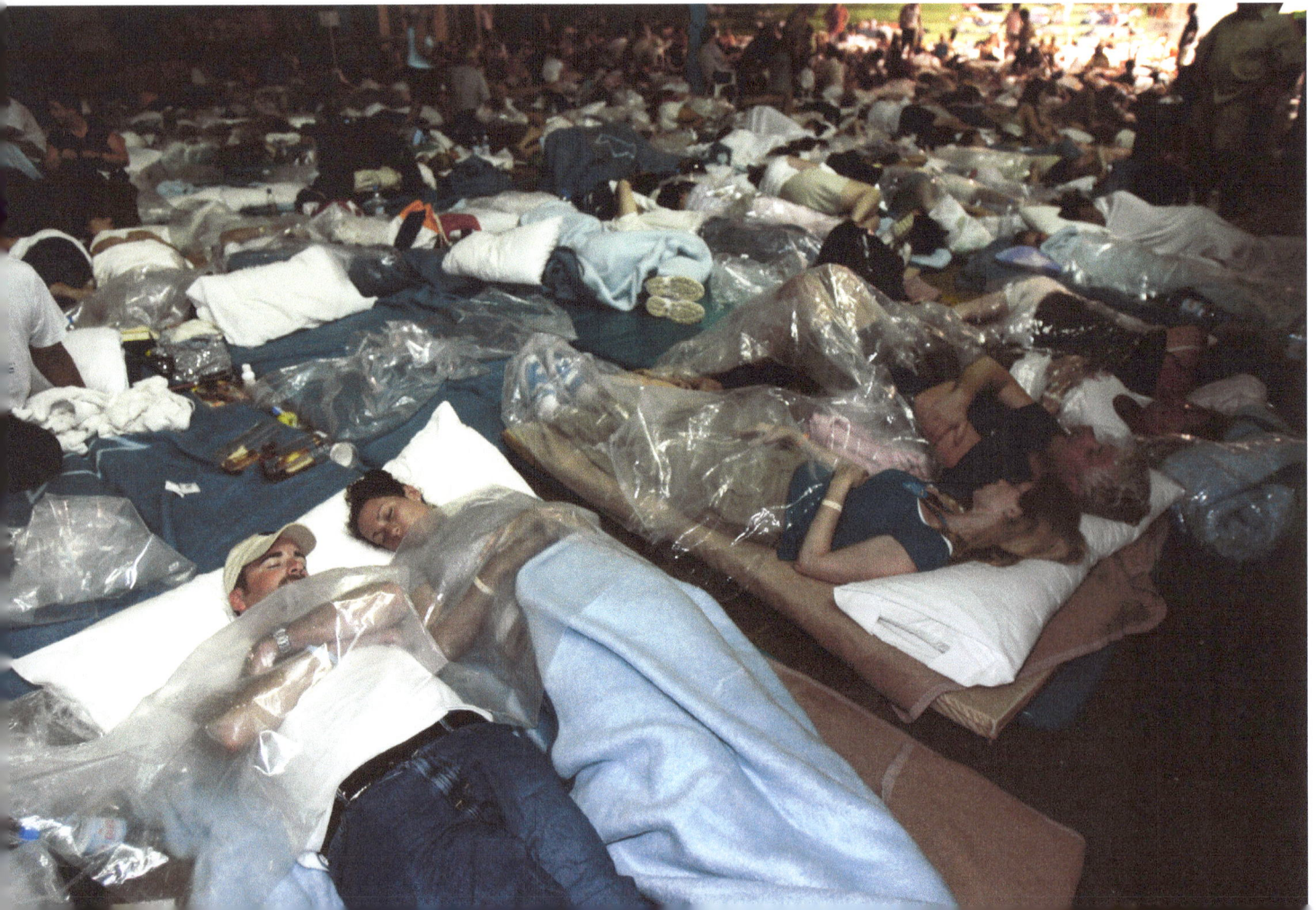

Record rainfall

Wilma formed as a tropical **depression** on Oct. 14, 2005, near the island of Grand Cayman. In just four days, it rapidly intensified from a tropical storm with winds of 69 miles (111 kilometers) per hour to a Category 5 hurricane. Wilma devastated the northeastern Yucatán. The World Meteorological Organization reported that Isla Mujeres, an island in the region, recorded the largest 24-hour rainfall ever in the Northern Hemisphere, at more than 64 inches (160 centimeters).

Havoc in Florida

As **meteorologists** predicted, Wilma weakened after it left the Yucatán. However, the hurricane strengthened unexpectedly in the warm waters of the Gulf of Mexico. The storm reached Category 3 before making **landfall** near Cape Romano, Florida, on October 25. South Florida residents were caught off guard. Winds of 125 miles (200 kilometers) per hour uprooted trees, leveled buildings, and knocked out power. More than 6 million people were left without electricity, and there was widespread property and crop damage.

RECORD STORM SEASON

The year 2005 was the busiest Atlantic storm season on record. There were an unprecedented 28 storms, including several that reached Category 5, including hurricanes Katrina, Rita, and Wilma. The year also marked the first time that forecasters ran out of names for storms from the regular chart of tropical system names. They resorted to using letters of the Greek alphabet to name a few storms. The 2005 season began with Tropical Storm Arlene in June and ended with Tropical Storm Zeta in late December.

A combination of satellite images shows Hurricane Wilma at different points along its path.

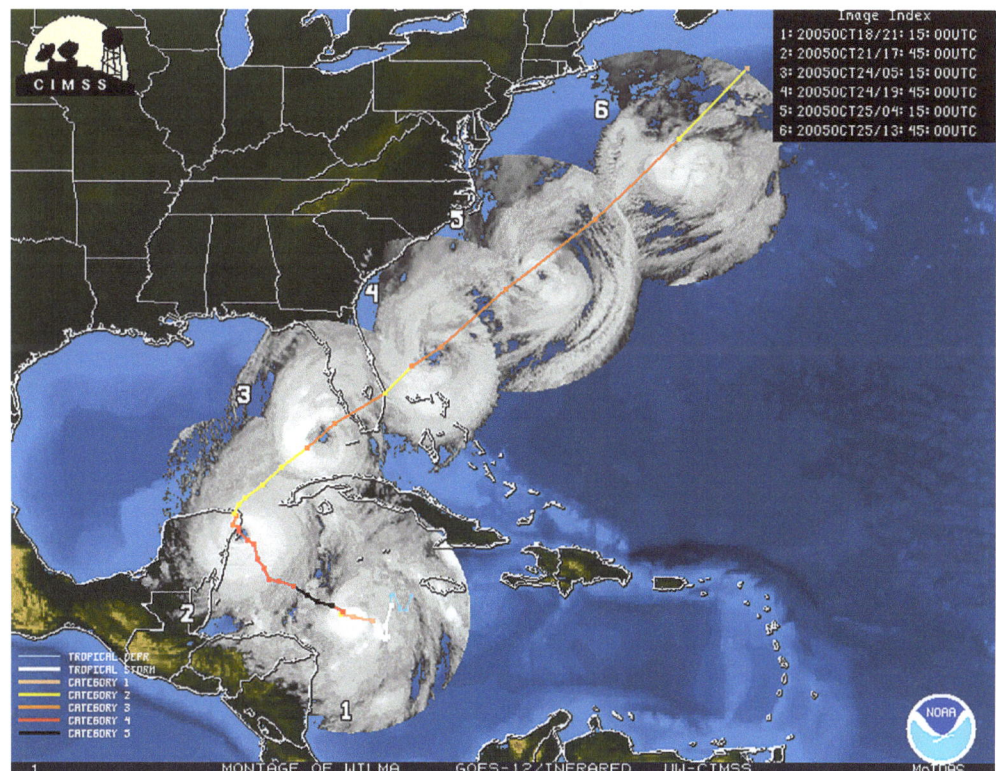

2017 ATLANTIC HURRICANE SEASON

The 2017 Atlantic hurricane season was one of the worst on record. A series of devastating storms from late August to early October left parts of the United States, Mexico, Nicaragua, and many Caribbean islands badly damaged.

Harvey

On August 17, a tropical storm developed north of the South American coast and was named Harvey. The storm encountered unfavorable conditions as it traveled west. However, it rapidly gained strength after crossing the Yucatan Peninsula in southern Mexico, becoming a hurricane on August 24. It continued to strengthen over the following hours, making landfall near Corpus Christi, Texas, in the evening of August 25. Weather systems parked over the United States prevented the hurricane from moving inland. Instead, it stalled over Houston, sucking moisture from the Gulf of Mexico and dumping rain onto land. A broad area of southeast Texas received 30 to 45 inches (75 to 115 centimeters) of rain in just a few days. So much rain fell that the National Weather Service had to modify the colors on its rainfall chart to map it all effectively. More than 90 people died in the storm. Harvey caused some $125 billion in damage, tying it with Katrina (see pages 22-25) as the costliest Atlantic hurricane of all time.

Hurricane Harvey flooded large portions of Houston, Texas, and the surrounding suburbs.

Residents wait in line at a Puerto Rico gas station after Hurricane Maria, when gasoline-powered generators were the only source of electric power for weeks.

Irma

On August 30, Tropical Storm Irma formed off the coast of Africa. For the next week, Irma gathered strength as it swirled westward across the Atlantic Ocean, reaching Category 5. Beginning September 6, it tore through the Caribbean Islands. Some 95 percent of buildings on the island of Barbuda were damaged or destroyed. The entire peninsula of Florida lay in Irma's path, prompting one of the largest evacuations in U.S. history. The hurricane made landfall on the Gulf Coast of Florida on September 10 as a strong Category 4. The damage to the state was not as severe as had been initially feared, but many islands in the Florida Keys were devastated. Over 100 people were killed throughout the Caribbean and the mainland United States. Hurricane Irma caused about $50 billion in damage.

Maria

The storm that would become Hurricane Maria formed in the central Atlantic on September 13. It strengthened to a Category 5 hurricane five days later, just before entering the Caribbean Sea. Maria struck the island of Dominica the next day, causing catastrophic damage and killing 65. The following day, it hit the U.S. territory of Puerto Rico. Puerto Rico was already struggling with an outdated, unreliable power grid. In addition, as an island, it could not draw power from the rest of the United States during the emergency. The direct hit from Maria resulted in a total loss of power throughout the island. The storm caused $90 billion in damage. The official death toll is 64, but most experts think hundreds or thousands of people died.

HURRICANE SANDY

A monster storm struck the Atlantic region shortly before Halloween in 2012. The storm, which formed over the Caribbean, was officially named **Hurricane** Sandy. But as it made its way along the East Coast of the United States, it collided with cold arctic air, producing a freakish storm nicknamed "Frankenstorm" and "Superstorm Sandy." Strong winds and flooding caused great destruction in the Caribbean Islands and along the U.S. coast, directly killing about 150 people. Cuba, Haiti, New York City, and coastal New Jersey were especially hard hit. Sandy also brought blizzard conditions to the Appalachian Mountains.

Development and path

Sandy began as a tropical **depression** over the southwestern Caribbean Sea on October 22. It increased to hurricane strength before making **landfall** near Bull Bay, Jamaica, on October 24. The storm then proceeded to Cuba, where it reached its maximum strength. On October 25, it made landfall west of Santiago de Cuba as a Category 3 hurricane, with wind speeds of 115 miles (185 kilometers) per hour. The storm then weakened as it traveled north, but it also grew in size. When Sandy made landfall near Brigantine, New Jersey, on October 29, it had become

Floodwaters from Hurricane Sandy pour into a New Jersey transit station.

an extratropical **cyclone**—that is, a cyclone outside the tropics—with winds of 80 miles (130 kilometers) per hour. Despite the weakened winds, the storm had a huge diameter of about 1,000 miles (1,600 kilometers). It continued on a path through southern New Jersey, northern Delaware, southern Pennsylvania, and northeastern Ohio, weakening gradually. The storm's remnants later merged with a low-pressure area in eastern Canada.

An unusual combination

Atlantic hurricanes usually do not travel into the inland United States. Instead, they tend to move eastward and out to sea. However, a weather system over the Atlantic Ocean prevented Sandy from taking this path, forcing the storm westward. The hurricane then joined with an arctic front from the northwest. Hurricane-strength winds and heavy rains battered the mid-Atlantic and New England coasts. Blizzard conditions hit the Appalachians, especially in West Virginia and North Carolina. To worsen matters, the storm made landfall in New Jersey while the ocean was at high **tide.** At Battery Park in southern Manhattan, a borough of New York City, the water level rose nearly 14 feet (4.3 meters) on October 29 due to the combination of high tide and **storm surge.**

A MONSTROUS IMPACT

Sandy caused about 160 direct and indirect deaths in the United States. Indirect deaths were due mainly to accidents caused by unsafe conditions following the storm. Sandy also brought historic devastation to New York City, where it flooded subway tunnels and left millions of people without electric power. Coastal areas of New Jersey also suffered great damage.

This satellite image, taken at the peak intensity of "Superstorm Sandy," shows the vast area affected by the storm.

STUDYING HURRICANES

Thousands of weather stations around the world constantly measure wind speed, **atmospheric pressure,** temperature, and **humidity.** The results are sent to special centers in the world's **hurricane** regions. These weather stations include the U.S. National Hurricane Center (NHC) in Miami, Florida, and the Central Pacific Center in Honolulu, Hawaii. There are also **tropical cyclone** centers in New Delhi, India; Réunion (an island in the Indian Ocean); Tokyo, Japan; and Fiji (an island in the Pacific Ocean). The Australian Bureau of Meteorology has **cyclone** centers in Brisbane, Darwin, and Perth; and New Zealand has a center in Wellington. Canada's hurricane center is located in Dartmouth, Nova Scotia.

Satellites and computers

Meteorologists use weather balloons, **radar,** satellites, and airplanes to study the weather and help forecast hurricanes. The balloons carry instruments for recording data in the upper

Hurricane Katrina roars toward the U.S. states of Louisiana and Texas on Aug. 28, 2005, in an image produced using the Tropical Rainfall Measuring Mission (TRMM) satellite. TRMM can peer beneath the clouds to measure a storm's rainfall. Red regions show the highest rainfall— at least 2 inches (5 centimeters) of rain per hour.

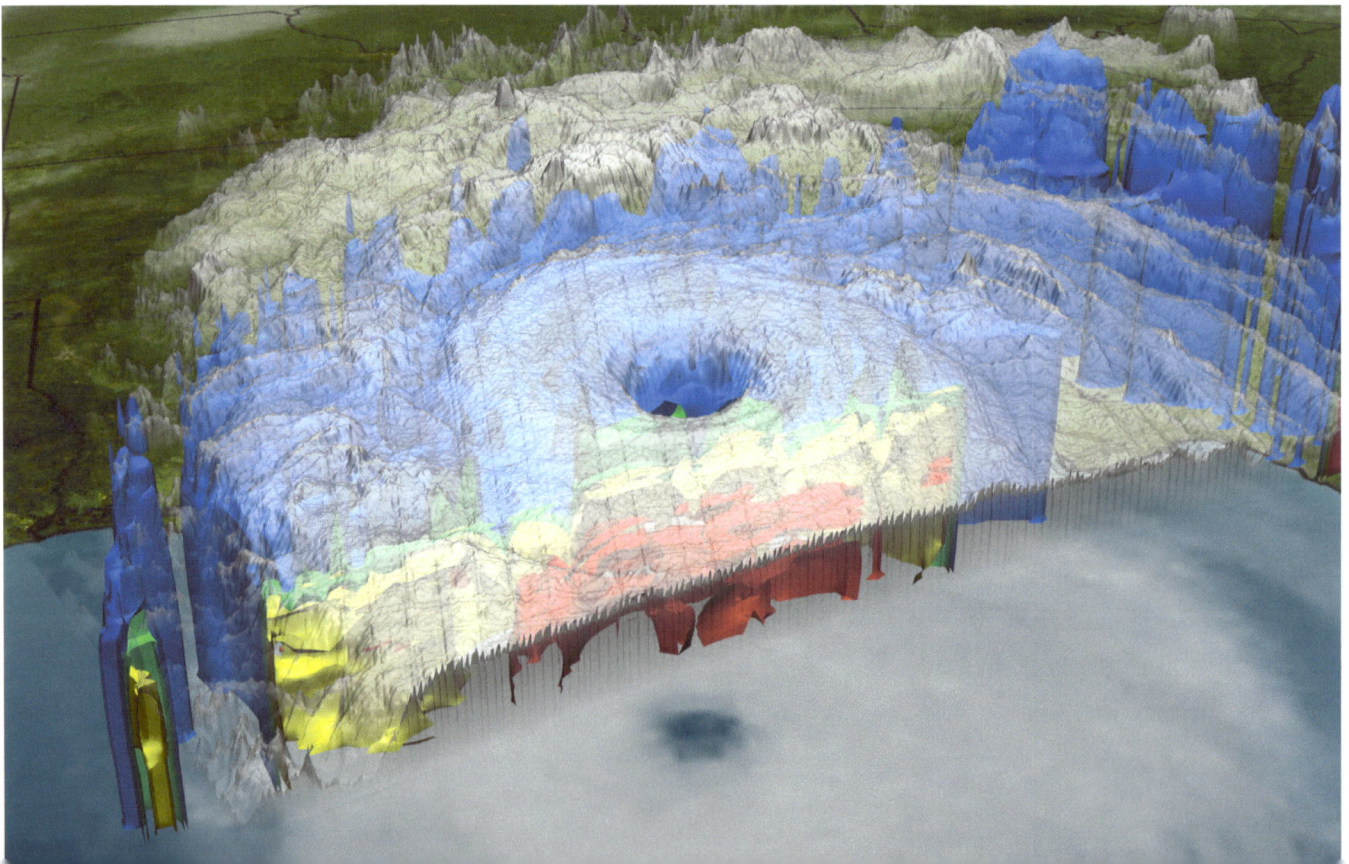

atmosphere, and radar is used to track storms. Geostationary Operational Environmental Satellites (GOES) monitor the weather on Earth 24 hours a day. They circle Earth in a geosynchronous (*JEE oh SIHNG kruh nuhs*) orbit. In such an orbit, a satellite's speed matches the planet's rotation, so that the satellite stays over a fixed position on Earth's surface. Each satellite is about 22,300 miles (35,900 kilometers) above the surface of Earth, high enough to give a wide view. The U.S. National Oceanic and Atmospheric Administration (NOAA) also has aircraft that fly right into and through hurricanes. The specially equipped planes are buffeted by howling winds, blinding rain, and hail, but they are able to drop instruments into the inner regions of storms. These instruments take measurements, which are radioed back to the aircraft, and the data are sent on to the National Hurricane Center by satellite.

At the hurricane centers, data from weather stations, satellites, and the instruments dropped by airplanes are fed into computers, and meteorologists use the information to predict how a storm will develop. The NHC keeps a continuous watch on hurricanes throughout the hurricane season, from May 15 in the eastern Pacific and June 1 in the Atlantic through the end of November.

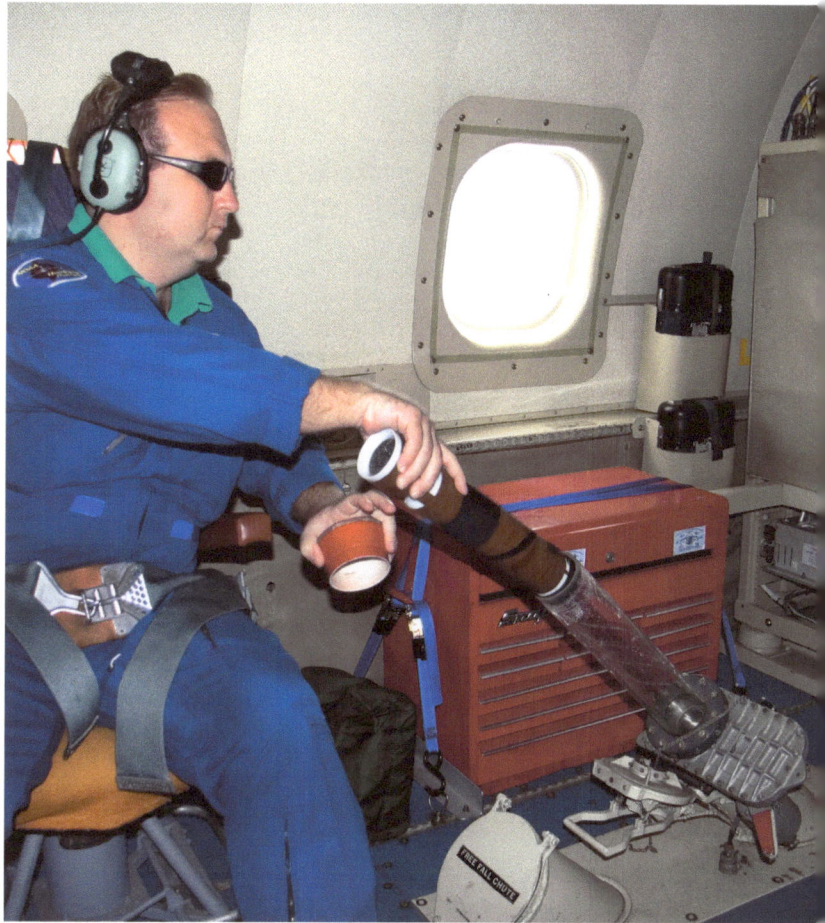

A researcher prepares to release a dropwindsonde into a hurricane.

THE DROPWINDSONDE

In the early 1970's, the National Center for Atmospheric Research (NCAR) developed the first dropwindsondes *(DRAHP wihnd sahndz),* small tubes containing instruments that could measure wind speeds and temperatures inside hurricanes. The dropwindsonde, or dropsonde for short, has a parachute at the top and is released from an airplane. In the early 1990's, NCAR developed dropwindsondes with Global Positioning System (GPS) capabilities.

OKEECHOBEE AND LONG ISLAND

The port of New London, Connecticut, in the northeastern United States, lies damaged by the 1938 hurricane known as the "Long Island Express."

In 1928 and 1938, two great **hurricanes** struck the United States. **Meteorologists** label both storms as Cape Verde (*kayp VURD*) hurricanes because they are thought to have originated near the Cape Verde Islands, off the west coast of Africa. Such storms become very intense because they move over a large expanse of warm ocean water before reaching land in the Caribbean or North America.

Lake surge

The Hurricane of 1928, often called the San Felipe Segundo in Puerto Rico and the Lake Okeechobee Hurricane in the United States, hit the Leeward Islands, Puerto Rico, and the Bahamas before it made **landfall** near Palm Beach, Florida, in the southeastern United States, on September 16. The storm then headed for Lake Okeechobee, the largest lake in the southern United States. Winds measuring about 140 miles (225 kilometers) per hour produced a **storm surge** on the lake. It washed over the 5-foot- (1.5-meter-) high **levee** of earth and rock at the lake's south end. Soon a flood up to 11 feet (3.3 meters) deep covered hundreds of thousands of acres of farmland, and houses were crushed or swept off their foundations. As the **eye** passed over the lake, another surge went in the opposite direction, flooding the lake's northern coast. The total number of deaths caused by the hurricane is unknown, but it may have exceeded 4,000, including the dead from the Caribbean and the Bahamas, where the storm first made landfall.

The "Long Island Express"

In September 1938, a Cape Verde hurricane turned north before reaching Florida. It began moving faster, and its forward speed reached more than 60 miles (100 kilometers) per hour as it passed east of North Carolina. This is about three times as fast as the usual forward motion of a hurricane, so the hurricane earned the nickname of the "Long Island Express." The storm struck Long Island, New York, at 3:30 p.m. on September 21, just before high **tide,** creating a storm surge of up to 12 feet (3.7 meters). In New York City, wind speeds of 120 miles (190 kilometers) per hour were recorded at the top of the Empire State Building. In nearby Central Park, wind speeds at ground level were measured at 60 miles (100 kilometers) per hour. The storm then moved on to New England, where the states of Connecticut and Rhode Island were especially hard hit. In downtown Providence, Rhode Island, a storm surge caused floodwaters to rise from a few inches or centimeters deep to waist-high in minutes, and the waters were soon 20 feet (6 meters) deep. The hurricane of 1938 caused nearly 700 deaths.

KEEPING THE LIGHT

Captain Arthur A. Small was the keeper of the lighthouse on Palmer's Island near New Bedford, Massachusetts. On Sept. 23, 1938, he dictated a letter to the superintendent of lighthouses, describing the events of two days earlier: "Keeper swept overboard, but by swimming underwater, made the station again. Mrs. Small, the keeper's wife, was seen by the keeper while he was overboard. She left the oil house where he had told her to stay and evidently she tried to launch a boat to save the keeper, but she was swept away and drowned. ... There is no shelter to be had at the station, except in the top of the tower. Keeper remained on duty until properly relieved. ... Keeper removed to St. Luke's Hospital suffering from exhaustion and exposure."

A house at West Hampton, Long Island, was torn from its foundations and blown onto the beach by the hurricane of 1938.

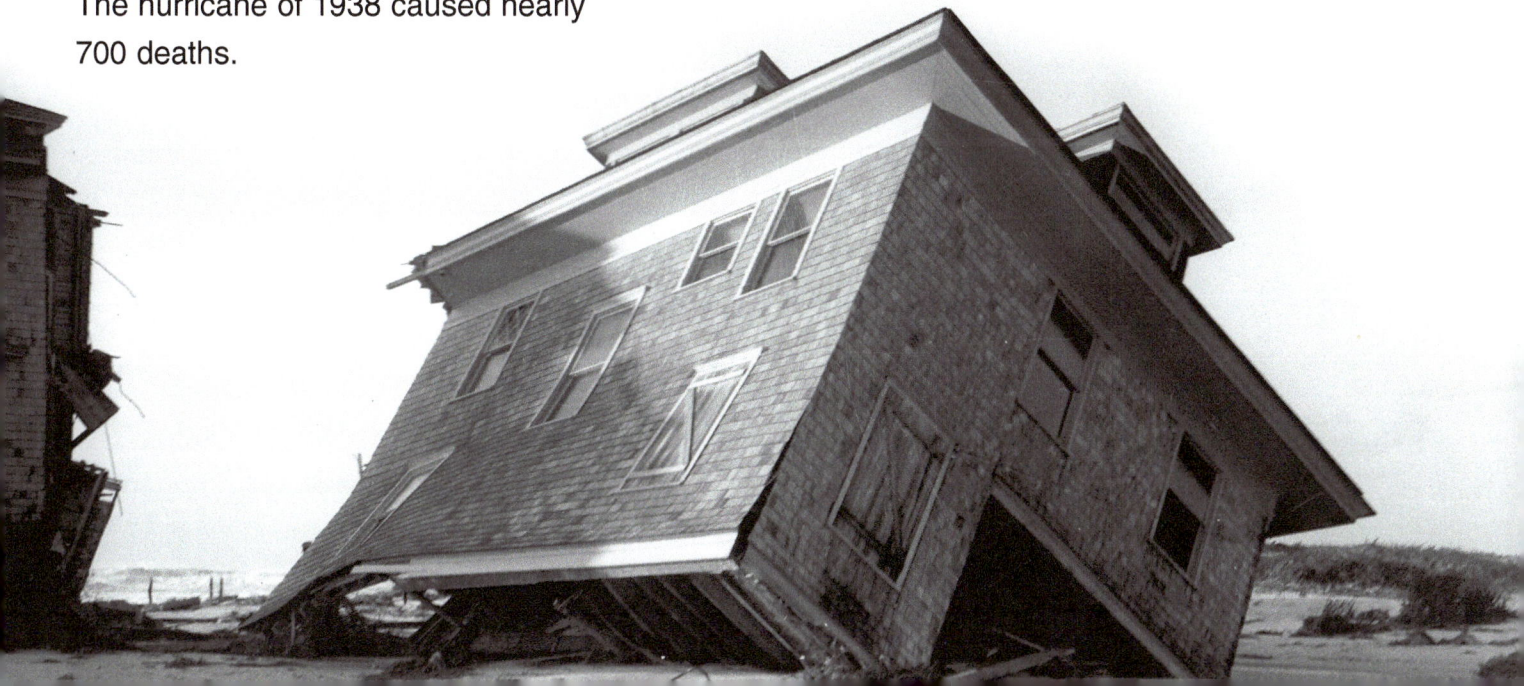

BIG WINDS OF THE PACIFIC

A worker repairs damaged power lines in Zhejiang Province, China, after Typhoon Saomai devastated the region in August 2006.

In the northwest region of the Pacific Ocean, **tropical cyclones** are called **typhoons.** Typhoons are very frequent. On average, 25 typhoons strike the region each year, usually between the months of June and November. Traveling west, like most tropical cyclones, typhoons threaten the coasts of the Pacific Islands, China, and Japan.

From the Pacific to the Indian Ocean

Some typhoons travel very long distances. In 1997, Typhoon Linda crossed from the Pacific Ocean all the way to the Indian Ocean. It formed in the Pacific on October 26, moved west over the Philippines as a tropical storm, and strengthened over the South China Sea. On November 2, Linda caused great damage to the southern tip of Vietnam as it reached typhoon strength (the same as **hurricane** strength; see page 11). It then went straight over the Malaysian peninsula and into the north Indian Ocean, where it weakened by November 10. This typhoon caused nearly 500 deaths. There was no time to warn people in small fishing boats off the coast of Vietnam about the coming storm, and many were lost at sea that day.

Super typhoons

In the southwest Pacific, typhoons with wind speeds of more than 150 miles (241 kilometers) per hour—approaching the strength of a Category 5 hurricane—are called **super typhoons.** In 2006, China was hit by its strongest super typhoon in half a century. When Typhoon Saomai *(sow my)* reached the Chinese coast on August 10, more than 1.7 million people were evacuated and a state of emergency declared. The government sent text messages, posted messages on the Internet, and broadcast warnings on television and radio to warn people to leave dangerous areas. Nevertheless, more than 450 people died in the storm.

TYPHON

Scholars do not know for sure the origin of the word *typhoon.* In ancient Greek mythology, Typhon *(TY fon)* was the son of the Earth goddess Gaia *(GAY uh).* He was a monster with a hundred serpent heads and fiery eyes. In ancient Greek, the god's name meant "whirlwind." The king of the gods, Zeus, attacked Typhon with thunderbolts and cast him into the underworld. From there, Typhon went on creating windstorms to cause disaster on Earth.

People in the Chinese city of Fuding are battered by severe winds and heavy rain as Typhoon Saomai makes landfall to the north.

JAPAN AND THE PHILIPPINES

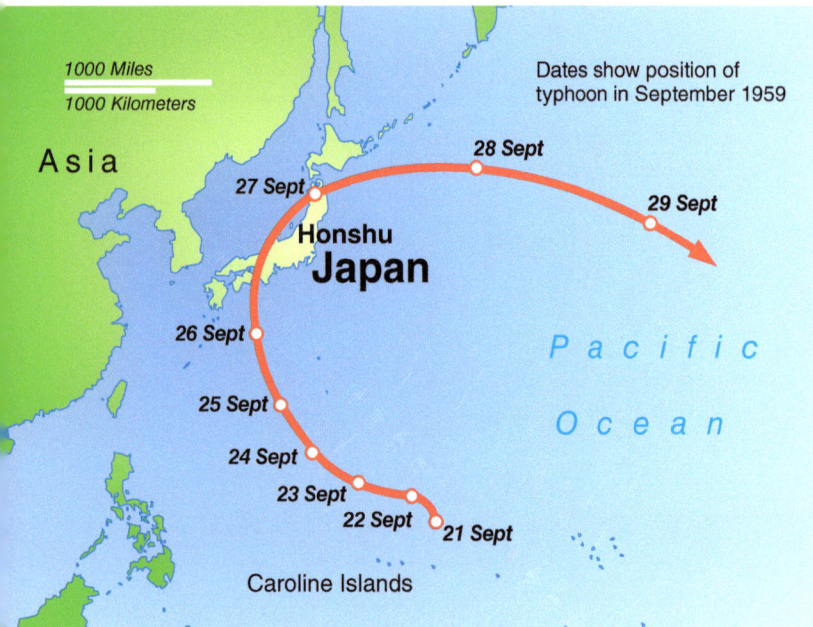

Most **typhoons** begin as areas of low **atmospheric pressure** over the Pacific Ocean near the **equator.** Some form over or near the Caroline Islands, a group of small islands north of New Guinea and east of the Philippines. Two of the region's deadliest storms started in that area, but they took very different paths thereafter. **Super Typhoon** Vera, called Isewan (*ee say wahn*) in Japan (see page 41 for an explanation of typhoon names), curved quickly toward the north and hit Japan. Tropical Storm Thelma, however, traveled almost directly west. Like many tropical storms, it crossed the Philippines and passed into the South China Sea.

Devastating Honshu

Five days after developing west of the Caroline Islands on Sept. 21, 1959, Vera made **landfall** at the south end of the largest Japanese island, Honshu. By then, Vera was an extremely violent super typhoon, and it headed northeast to run the length of the island's west coast. A **storm surge** of up to 11 feet (3.5 meters) and heavy rainfall caused flooding and **landslides.** The storm left behind a death toll of around 5,000, with nearly 39,000 people injured. It destroyed large areas of crops, as well as roads and railways, and left 1.5 million people homeless.

The path of Super Typhoon Vera in 1959, which had a top wind speed of 190 miles (305 kilometers) per hour and traveled a distance of 2,898 miles (4,664 kilometers).

Tropical Storm Thelma's path in 1991. Although much less violent than Vera, Thelma caused more deaths.

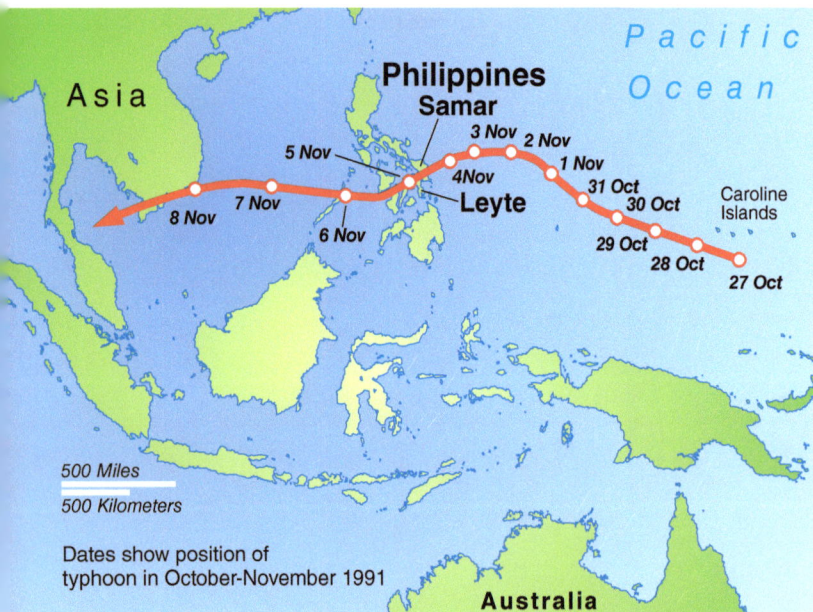

Thelma

Tropical Storm Thelma, called Uring in the Philippines, began at the east end of the Caroline Islands. By the time it made landfall on the Philippine island of Samar, on Nov. 4, 1991, its winds had reached a peak speed of 50 miles (80 kilometers) per hour. Yet, despite the fact that Thelma never developed into a full-blown typhoon, the storm's effects were devastating. It dropped more than 6 inches (15 centimeters) of rain in 24 hours on the island of Leyte, causing floods, **mudslides,** and **landslides.** About 6,000 people died.

A family surveys the damage after Thelma, a 1991 tropical storm, dropped enormous amounts of rain on Leyte Island in the Philippines. Subsequent flooding and mudslides killed as many as 6,000 people.

MASSIVE MUDSLIDES

The town of Ormoc, on Leyte Island, was badly damaged by Tropical Storm Thelma. Three-quarters of the town was destroyed when it was hit by mudslides, and more than 3,000 people died. Later, an investigation of the storm determined that loggers had acted illegally and cleared more trees from the area than their permits allowed. The loss of so many trees made the slopes of the hills around the town unstable. When the torrential rains from Thelma washed loose mud and soil down the slopes, it caused massive mudslides. **Deforestation** (*dee FAWR uh STAY shuhn*) has caused similar problems in other parts of the world.

TYPHOON COBRA

Typhoons, cyclones, and hurricanes are deadly hazards to ships at sea. In one famous incident during World War II (1939-1945), a sudden typhoon in the Philippine Sea hit a group of U.S. warships. The ships were about 300 miles (500 kilometers) east of the island of Luzon, the nearest land.

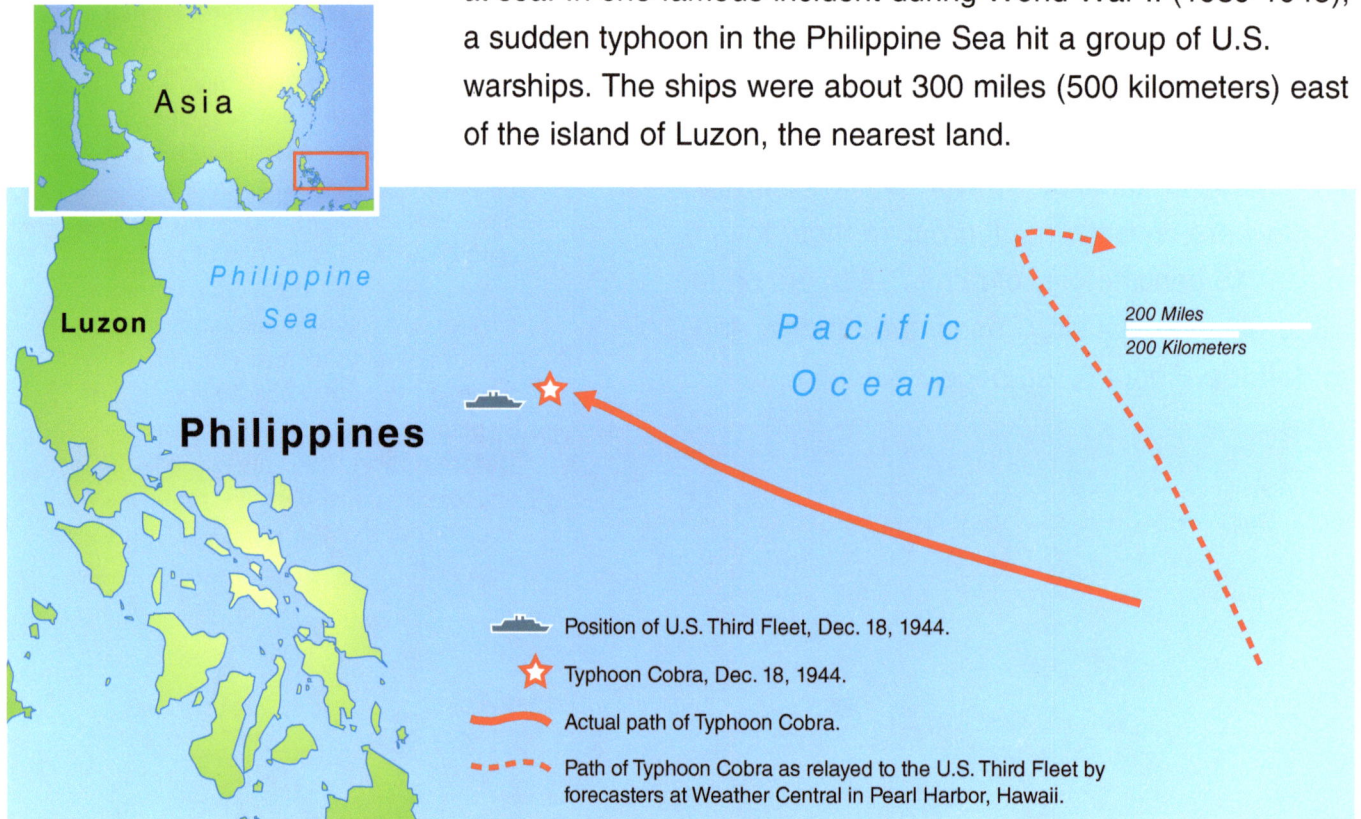

Asia

Luzon

Philippine Sea

Philippines

Pacific Ocean

200 Miles
200 Kilometers

Position of U.S. Third Fleet, Dec. 18, 1944.

Typhoon Cobra, Dec. 18, 1944.

Actual path of Typhoon Cobra.

Path of Typhoon Cobra as relayed to the U.S. Third Fleet by forecasters at Weather Central in Pearl Harbor, Hawaii.

The path that forecasters predicted the 1944 Typhoon Cobra would take, compared with the storm's actual path. Cobra sank three U.S. Navy destroyers, damaged several other ships, and caused the death of some 790 men.

Taken by surprise

A task force of ships from the U.S. Third Fleet was made up of 13 aircraft carriers, 8 battleships, 15 cruisers, and about 50 destroyers. This force had just completed raids on Japanese airfields. Suddenly, on Dec. 18, 1944, Typhoon Cobra surprised the force while some ships were refueling. One naval officer described the scene: "By 1300 [1 p.m.] we must have passed through the center for there was a momentary lull—the seas hit us from all directions and the ship was racked and twisted—but she survived. The respite from the wind was only a matter of minutes, then it howled, whined and finally got back to shrieking again."

Destroyers sunk

Three destroyers sank in the typhoon, which had gusts of up to 115 miles (185 kilometers) per hour. There was severe damage to other ships, and about 150 aircraft were wrecked on the carriers. About 790 men were killed, with fewer than 100 survivors from the lost destroyers. The commander of the fleet, Admiral William Halsey, later wrote: "No one who has not been through a typhoon can conceive its fury. The 70-foot [20-meter] seas smash you. The rain blinds you. ... The typhoon tossed our enormous ship the *Missouri* as if she were only a canoe."

One of the men serving with the Third Fleet was Gerald Ford (1913-2006), who in 1974 would become the 38th president of the United States. When the typhoon hit the aircraft carrier U.S.S. *Monterey,* planes in the ship's hangar deck were tossed about, spilling fuel that caused a deadly fire. As an officer on the *Monterey,* Ford went down to the hangar deck to assess the fire, battling smoke and pitching jolts from the storm. His heroism helped to save many lives.

TYPHOON NAMES

Cobra was an unofficial name for the typhoon because official names were not yet in use in 1944. After World War II, the Guam Joint Typhoon Warning Center started to use lists of names for Pacific storms. Members of the Typhoon Committee of the World Meteorological Organization (WMO) began naming these storms in 2000. The WMO gathered 10 names from each of the 14 countries affected by typhoons and put the names into lists. At the beginning of each typhoon season, the WMO continues with the next name on the list. In addition, the Philippines and Australia have their own name lists.

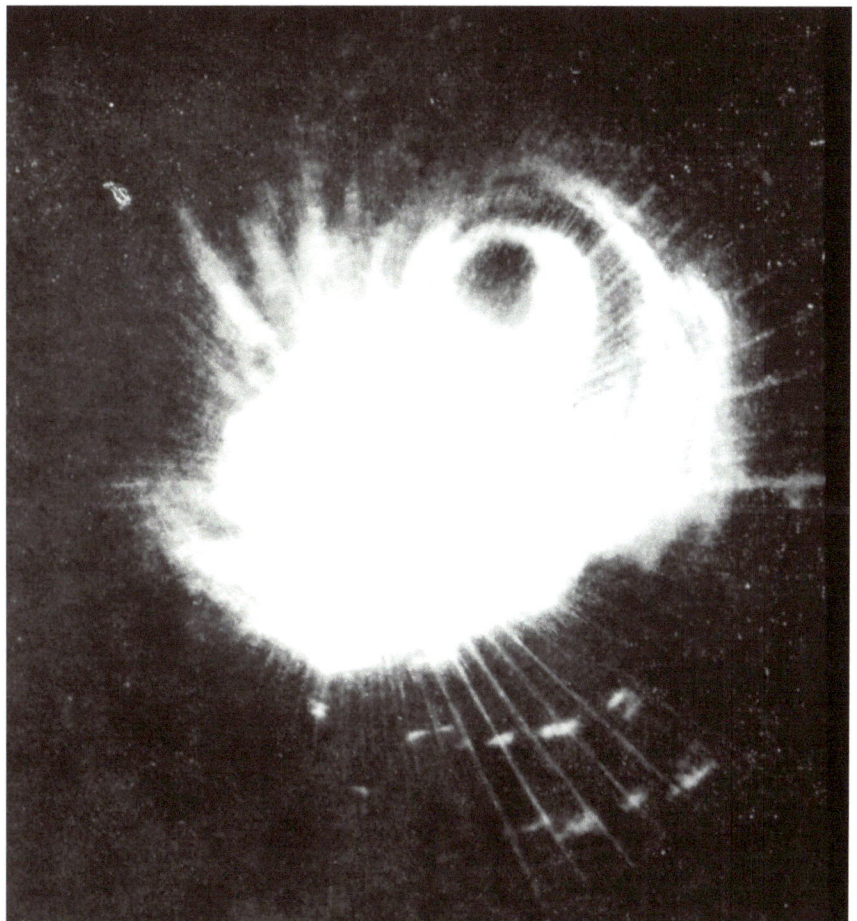

Typhoon Cobra as it appeared on a radar screen from a U.S. Navy ship in December 1944. Cobra was only the second tropical storm ever to be observed on radar.

STORMS OFF THE INDIAN OCEAN

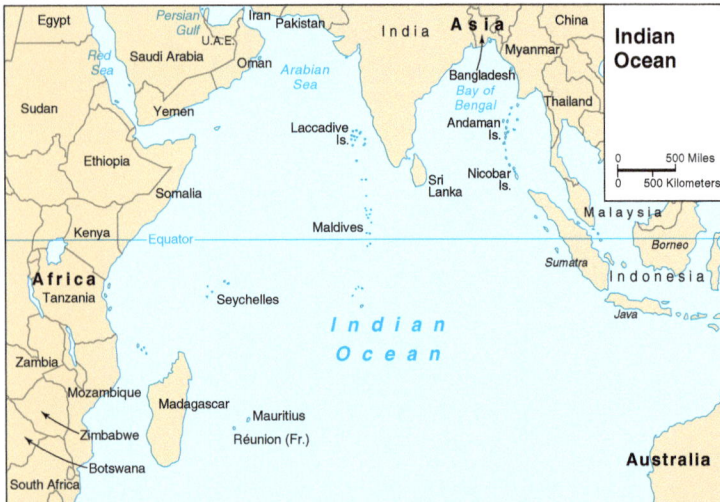

On average, about 17 severe storms form each year over the Indian Ocean.

An aerial shows devastation caused by a cyclone that hit the state of Orissa, in eastern India, in November 1999.

In the Indian Ocean region, severe storms are called **tropical cyclones,** or simply **cyclones.** On average, 12 cyclones form every year just south of the **equator.** They move southwest across the Indian Ocean toward Madagascar and the coast of Africa. Each year, about five severe storms form just north of the equator. They move northwest toward Sri Lanka and India. Some travel to the west of the Indian subcontinent. Others stay east and head for the Bay of Bengal, where they often cause destruction in the coastal country of Bangladesh (see pages 44–45).

Blown by monsoon winds

A **monsoon** is a wind that reverses itself seasonally, especially one that blows over the Indian Ocean and surrounding land areas. The Indian monsoon blows continually from the southwest from June through September, bringing heavy rains to southern and southeastern Asia. Then it changes direction to blow from the northeast from December to March. Cyclones in this region are not necessarily caused by the monsoon, but the warm monsoon winds drive the cyclones northward across the Indian Ocean.

From Asia to Africa

In February 2000, an area of low **atmospheric pressure** formed near the Indonesian island of Java and headed west. Named Leon by the Australian **meteorology** bureau, the pressure system took two weeks to travel nearly 4,000 miles (6,400 kilometers) across the Indian Ocean. On the way, it was renamed Tropical Cyclone Eline by the Mauritius bureau. When it struck the east coast of Madagascar, it left tens of thousands homeless. The storm passed over the island and went on to cause destruction and flooding in southern Africa. This was particularly devastating, as certain areas of Botswana, Namibia, South Africa, Swaziland, and Zimbabwe had already experienced torrential rainfalls with flooding that February. Mozambique was especially hard hit. The effect of Eline on top of already heavy rains caused some of the worst flooding seen in southern Africa for 50 years, leaving an estimated 800,000 people homeless or otherwise in danger. In Mozambique, most of the country's farmland was submerged, crops were lost, and roads and bridges were washed away. In hilly regions, many fled to the hilltops and were then stranded there. They waited for days for small boats or other transport to arrive to rescue them. Such diseases as cholera, malaria, and meningitis spread throughout the area.

RECORD RAINFALL

Réunion—an overseas department, or administrative district, of France—lies about 400 miles (640 kilometers) east of Madagascar. The island has been hit by many tropical cyclones, which often drop huge amounts of rainfall. Réunion holds the world record for the largest rainfalls from cyclones. In 1966, Tropical Cyclone Denise caused 71.8 inches (183 centimeters) of rain to fall on the island in one day. Fourteen years later, Tropical Cyclone Hyacinthe dropped 223.5 inches (567.8 centimeters) over a 10-day period.

A boy searches through the rubble of an orphanage after a cyclone in 2008 devastated the coast of Myanmar, leaving more than 78,000 people dead.

BANGLADESH

Three major rivers—the Jamuna *(JUH muh nuh),* or Brahmaputra *(BRAH muh POO truh);* Padma *(PUHD mah),* or Ganges *(GAN jeez);* and Meghna *(MAYG nuh)*—flow across the flat plains that cover most of the country of Bangladesh. The rivers join to form the huge Ganges **delta.** Most of Bangladesh lies less than 50 feet (15 meters) above sea level, and the rivers often overflow during the rainy season. The floods are made much worse when **tropical cyclones** strike, which they often do in September, at the end of the **monsoon** season. Tropical cyclones cause huge **storm surges** to rush in from the Bay of Bengal. According to the Asian Disaster Reduction Center: "Only 5 percent of cyclones form in the Bay of Bengal, but loss of lives and property [there] is about 85 percent of the global total." In 1970, a severe storm that formed in the Indian Ocean caused one of the worst natural disasters in history.

Heading north

On Nov. 7, 1970, an area of low **atmospheric pressure** formed in the Indian Ocean between India and the Andaman Islands. During the night of November 12 and the morning of November 13, the

In the Bay of Bengal, the northern part of the Indian Ocean, in 1970, survivors wait for rescue after a tropical cyclone caused major flooding in the region.

developing cyclone headed north into the Bay of Bengal and made landfall over Bangladesh (which was then part of Pakistan and called East Pakistan). Winds of up to 139 miles (224 kilometers) per hour caused an extremely high storm surge of 20 to 33 feet (6 to 10 meters), which flooded a large part of southern East Pakistan. Many people were asleep when their homes were flooded or swept away.

Deadliest tropical cyclone

It is estimated that about 266,000 people died in the floods, but many people believe that the real figure may be as high as 500,000 dead or even more. There were several reasons this storm was so deadly: (1) People received very little warning of the approaching storm. East Pakistan was part of a poor country with little money to invest in safety measures, so no storm-warning systems had been set up. (2) There were few flood barriers to hold back the surging **tide** and no cyclone shelters. (3) After the storm struck, it was very difficult to get aid to the devastated areas quickly enough.

DEALING WITH DISASTER

After the 1970 disaster, East Pakistan became Bangladesh, and the new government set up the Storm Warning Centre and built hundreds of storm shelters. In 1991, another cyclone killed about 138,000 Bangladeshis. Since then, warning systems have been improved, and people are better informed about what to do when a warning is issued. More recent cyclones and storm surges in Asia have killed far fewer people than did previous storms.

Survivors of the typhoon and storm surge that hit East Pakistan in 1970 scramble for food dropped by U.S. Army relief helicopters.

THE DESTRUCTION OF DARWIN

About 15 **tropical cyclones** develop each year in the southwest region of the Pacific Ocean (see map on page 5). Many head toward the northern coast of Australia, and some curve eastward to Fiji and other island groups. One of the most destructive storms pounded the Australian port of Darwin on Christmas Day, 1974.

Small but deadly

On Dec. 20, 1974, **meteorologists** noted that a tropical **depression** was forming in the Arafura Sea, between Australia and New Guinea. Within 24 hours, the growing storm was upgraded to a tropical cyclone and named Tracy. For days, however, no one in Darwin was concerned, because the storm was heading well to the west of the city. Further, about 10 days before Tracy, a cyclone had been predicted to reach Darwin, but instead turned and left the city unaffected. For these reasons, most residents of Darwin were not anxious about Tracy. On the evening of December 22, the local radio news reported: "Cyclone Tracy poses no immediate threat to Darwin." It was a small cyclone, with gale-force winds extending about 25 miles (40 kilometers) from the center and an **eye** about 7 miles (11 kilometers) across.

The path followed by Cyclone Tracy in December 1974.

Trail of destruction

Then the unexpected happened. On December 24, Tracy passed Bathurst Island, off the north coast. Suddenly the storm made a left turn and headed straight for Darwin. It passed over Darwin's port in the early hours of December 25. There was a **storm surge** of up to 5 feet (1.6 meters) in the harbor, massive rainfall of 10 inches (26 centimeters) in 12 hours, and terrifying winds of 135 miles (220 kilometers) or more per hour. More than three-fourths of Darwin's houses were destroyed, 65 people were killed, and about 790 others were injured. Ships in the harbor trying to ride out the storm foundered, and many of them sank. The cyclone left more than 20,000 people homeless in Darwin.

Houses in Darwin were ripped apart or simply blown down by Cyclone Tracy's high winds.

CHANGING CYCLONE WARNINGS

■ *Cyclone Warning 15 issued by Darwin Tropical Cyclone Warning Centre at 10 AM 12/24/74:* At 9 AM severe tropical cyclone Tracy was centred 115 km [71 miles] WNW of Darwin and moving south at 4 km/h [2.5 mph]. The centre is expected to be 100 km west of Darwin at 9 PM today. Very destructive winds of 120 km/h [74 mph] with gusts to 150 km/h [93 mph] are expected to continue on Bathurst Island today.

■ *Warning 16 at 12:30 PM 12/24/74:* Very destructive winds ... are expected in the Darwin area tonight and tomorrow.

■ *Auxiliary Warning at 2:30 AM 12/25/74: ...* 18 km [11 miles] west north west of Darwin moving east southeast at 6 km/h [4 mph]. The eye of the storm is expected to move over Darwin soon.

TYPHOON HAIYAN

The Philippines was battered by the strongest **super typhoon** in recorded history in 2013. **Typhoon** Haiyan *(HY an),* known as Typhoon Yolanda in the Philippines, was the deadliest and most destructive natural disaster to hit the Pacific island nation. The typhoon slammed into the central coastal areas on November 8, packing sustained winds at 195 miles (315 kilometers) per hour, with gusts above 225 miles (360 kilometers) per hour. According to meteorologists, the storm remained at peak Category 5 intensity for an incredible 48 straight hours. It leveled towns and cities in its path and killed more than 6,000 people.

Survivors of Typhoon Haiyan sift through wreckage in Leyte province. The storm washed ashore a huge cargo ship.

Hardest hit islands

Typhoon Haiyan began as a tropical **depression** near the Caroline Islands, in the western Pacific Ocean, on November 3. It pushed west-northwest, rapidly gaining strength and speed until it became a full-blown typhoon. When it entered the Philippine Sea a few days later, Haiyan's winds had grown to super typhoon intensity—the strongest rating, at more than 150 miles (240 kilometers) per hour. The storm made **landfall** on Samar Island and soon hit nearby Leyte Island. Its heavy rains and violent winds kicked up a 20-foot (6-meter) **storm surge** that caused some of the greatest devastation and loss of life on the islands. Many people drowned or were crushed by collapsing buildings. More than 1 million were left homeless.

Vietnam evacuation

Barreling out of the Philippines, Typhoon Haiyan headed toward Vietnam, forcing the evacuation of hundreds of thousands of people. However, by the time Haiyan made landfall in Vietnam's northern province of Quang Ninh, it had weakened significantly and was downgraded to a tropical storm. Haiyan finally died out over central China.

The map below shows the path of Typhoon Haiyan through the Philippines.

SURVIVING A HURRICANE

In the past, people often had little warning of an approaching **hurricane.** Today, in most parts of the world, forecasters are able to give up-to-date information on radio, television, and the Internet. People who live in high-risk areas should know what action to take as soon as they receive a warning. They should be ready to leave their home and the area quickly.

Before the event

According to the U.S. National Hurricane Center, people should have an emergency plan ready. They should check flashlights and radios, buy food that will keep without refrigeration, store drinking water, and have a disaster supply kit. They should know which is the safest area in their home (usually a small, interior, ground-floor room) and learn the location of official public shelters. They should also have a supply of plywood or other material to board up windows. Some people buy their own private hurricane shelters, which are small, windowless,

Windows are boarded up to reduce damage and the danger of flying glass.

boxlike rooms that can be anchored to the floor inside the house. Other shelters are built partly or completely underground. Everyone must take any hurricane warning seriously. Sometimes there are false alarms, but they may be followed by a genuine emergency.

During the storm

If no evacuation is ordered for your area and you decide to stay at home, you should only do so if your house is sturdy and on high ground. Stay away from windows and doors, close all interior doors, and remain in the safest area of your home. If winds are very strong, lie on the floor beneath a table or other sturdy object. Remember that if the **eye** of the hurricane passes over you, it will seem like the storm is over. But suddenly the winds will return to hurricane force, only from the other direction. Listen for flood warnings, and be ready to move to higher ground at a moment's notice.

HURRICANE WATCH AND WARNING

The U.S. National Oceanic and Atmospheric Administration (NOAA) explains the difference between a hurricane watch and a hurricane warning this way: "*A hurricane watch* indicates the possibility that you could experience hurricane conditions within 36 hours. This watch should trigger your family's disaster plan, and protective measures should be initiated, especially those actions that require extra time, such as securing a boat, leaving a barrier island, etc. A *hurricane warning* indicates that sustained winds of at least 74 mph [119 kph] are expected within 24 hours or less. Once this warning has been issued, your family should be in the process of completing protective actions and deciding the safest location to be during the storm."

Sandbags give beachfront houses some protection against storm surges and floods.

MAKE YOUR OWN BAROMETER

A **barometer** measures **atmospheric pressure** (see page 9).

Equipment

- A tall, wide-mouthed, clear plastic bottle
- A dish or small bowl
- 2 thin pieces of wood (such as long matchsticks)
- Masking tape and a pen

1. Fill the bottle with water. Over a sink or bowl, press the dish to the top of the bottle and then turn it upside down. Let some of the water spill out into the dish.

2. Tilt the bottle to let some air bubble up. The bottle should end up about two-thirds full of water.

3. Slip the pieces of wood under the bottle to lift it off the base of the dish.

4. Stick a strip of masking tape to the side of the bottle and mark the water level.

Set the barometer in a safe place and watch it for several weeks. When the atmospheric pressure increases, the water level will rise in your barometer. When the pressure falls, the level will drop. Mark the level every day and keep a record. The reading will fall when stormy weather is approaching.

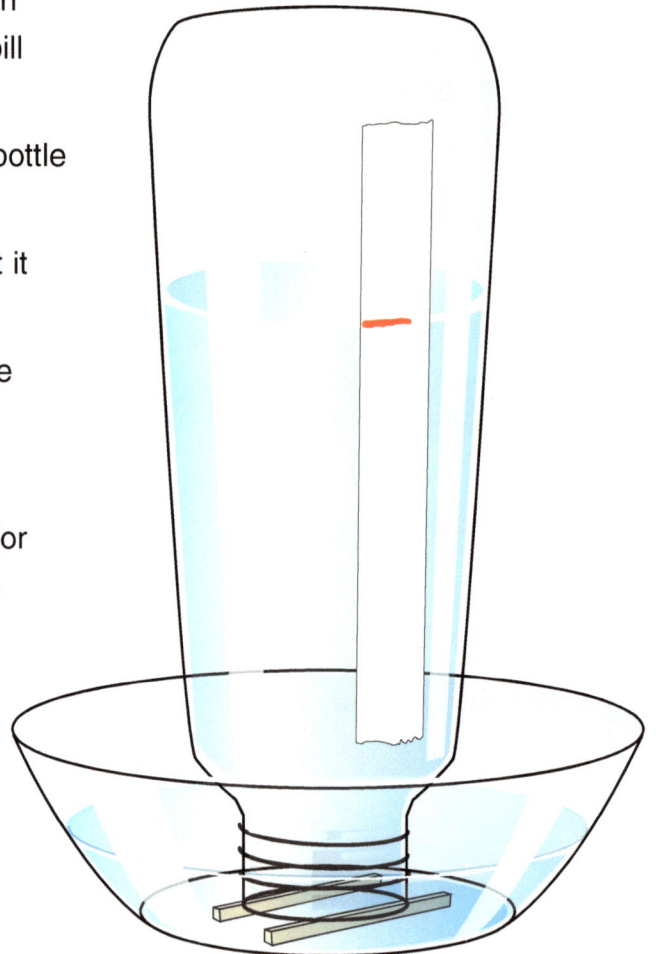

atmospheric pressure The weight of the air pressing down on Earth's surface.

barometer An instrument that measures the pressure of the atmosphere.

cumulonimbus A massive, vertical, cloud formation with peaks that sometimes resemble high mountains; such clouds may reach heights as great as 60,000 feet (18,000 meters).

cyclone A violent, swirling windstorm or the weather systems in which these windstorms form.

debris Rubble, broken objects, and other damaged material.

deforestation The clearing and removal of trees and forests.

delta The fan-shaped area at the mouth of some rivers, where the main flow splits into smaller channels.

depression Area of low atmospheric pressure.

equator An imaginary line around the middle of Earth, halfway between the North and South poles.

evaporate To change from a liquid into vapor.

eye The relatively calm area at the center of a cyclone or a hurricane.

eyewall A ring of thick clouds that spiral around the eye of a hurricane, carrying the strongest winds and rain.

gale A very strong wind; technically, any wind with a velocity of 32 to 63 miles (51 to 101 kilometers) per hour.

humidity The amount of moisture (water vapor) in the air.

hurricane A tropical storm over the North Atlantic Ocean, the Caribbean Sea, the Gulf of Mexico, or the Northeast Pacific Ocean.

landfall When the center of the eye of a hurricane first passes over a coastline. Note, when a hurricane makes landfall, half of the storm has already passed over the region.

landslide A mass of soil and rock that slides down a slope.

levee An embankment or wall built to prevent flooding.

mercury A silver-colored metallic element.

meteorologist A scientist who studies and forecasts the weather.

meteorology The study of Earth's atmosphere and the variations in atmospheric conditions that produce weather.

millibar (mb) A measure of atmospheric pressure (one-thousandth of a bar). Standard atmospheric pressure at sea level is about 1,013 mb.

monsoon A wind that reverses itself seasonally, especially the one that blows across the Indian Ocean and surrounding land areas.

mudslide Landslides of wet soil, or mud.

Northern Hemisphere The half of Earth that is north of the equator.

radar An electronic instrument used for determining the distance, direction, and speed of objects by the reflection of radio waves. Radar allows weather forecasters to locate areas of rain or snow and track the motion of air in a weather system.

rainband A band of cloud and rain that swirls around the eye of a hurricane.

Southern Hemisphere The half of Earth that is south of the equator.

storm surge A rapid rise in sea level produced when winds drive ocean waters ashore.

storm tide An especially high tide caused by stormy winds.

super typhoon A typhoon with wind speeds of 150 miles (242 kilometers) per hour or more—equivalent to a Category 4 hurricane.

tide The rise and fall of water in the ocean, caused primarily by variations in the gravitational pull of the moon and sun on different parts of Earth's surface.

trade wind A strong wind that blows toward the equator from the northeast or southeast.

tropical cyclone A large, powerful storm that forms over tropical waters.

tropics Regions of Earth that lie within about 1,600 miles (2,600 kilometers) north and south of the equator.

typhoon A tropical storm in the Northwest Pacific Ocean.

water vapor A gas formed by heating water.

weather system A particular set of weather conditions in Earth's atmosphere, which affects a certain area or region for a period of time.

Western Hemisphere The half of the world that includes North and South America.

BOOKS

Detecting Hurricanes, by Samantha S. Bell, Focus Readers, 2017.

Divine Wind: The History and Science of Hurricanes, by Kerry Emanuel, Oxford University Press, 2005.

Hurricane Force: In the Path of America's Deadliest Storms, by Joseph B. Treaster, Kingfisher, 2007.

Hurricane Katrina: Aftermath of Disaster, by Barb Palser, Compass Point Books, 2007.

Hurricane: Perspectives on Storm Disasters, by Andrew Langley, Heinemann-Raintree, 2014.

The Storm of the Century, by Al Roker, HarperCollins, 2015.

WEBSITES

http://www.noaa.gov/resource-collections/hurricanes

http://news.bbc.co.uk/2/hi/science/nature/4588149.stm

https://www.ready.gov/kids/know-the-facts/hurricanes

http://www.nationalgeographic.com/environment/natural-disasters/hurricanes/

http://www.nhc.noaa.gov

http://www.prh.noaa.gov/cphc/pages/FAQ/Tropical_Cyclone_Records.php

http://www.solar.ifa.hawaii.edu/Tropical/tropical.html

http://www.wmo.int/pages/prog/www/tcp/index_en.html

INDEX

www.ingramcontent.com/pod-product-compliance
Lightning Source LLC
Chambersburg PA
CBHW052042190326
41519CB00003BA/256

9 780716 694816